Global Food Security

CHALLENGES FOR THE FOOD
AND AGRICULTURAL SYSTEM

OECD

BETTER POLICIES FOR BETTER LIVES

This work is published on the responsibility of the Secretary-General of the OECD. The opinions expressed and arguments employed herein do not necessarily reflect the official views of the Organisation or of the governments of its member countries.

This document and any map included herein are without prejudice to the status of or sovereignty over any territory, to the delimitation of international frontiers and boundaries and to the name of any territory, city or area.

Please cite this publication as:
OECD (2013), *Global Food Security: Challenges for the Food and Agricultural System*, OECD Publishing.
http://dx.doi.org/10.1787/9789264195363-en

ISBN 978-92-64-19534-9 (print)
ISBN 978-92-64-19536-3 (PDF)

The statistical data for Israel are supplied by and under the responsibility of the relevant Israeli authorities. The use of such data by the OECD is without prejudice to the status of the Golan Heights, East Jerusalem and Israeli settlements in the West Bank under the terms of international law.

Photo credits: Cover © iStockphoto/Thinkstock.

Corrigenda to OECD publications may be found on line at: *www.oecd.org/publishing/corrigenda*.

Foreword

This study considers how changes to the world's food and agriculture system can contribute to improvements in food security in developing countries. It takes stock of a range of existing OECD work, including that undertaken with other international organisations, in particular for the G20, and places that work in the context of wider analysis both by international organisations and in academia. The purpose is to distil the main priorities for ensuring long-term global food security. The policy recommendations seek to improve the coherence of OECD countries' policies and contribute to multilateral initiatives, such as those pursued through the G20. More widely, the study seeks to contribute to the global debate on issues pertaining to global food security.

The report draws on a wide range of work at OECD. The synthesis was co-ordinated by Jonathan Brooks, and includes contributions from Claire Delpeuch, Professor Alan Matthews and Gloria Solano Hermosilla. The draft has benefited from the input of several colleagues within the OECD Secretariat and from Steve Wiggins at the Overseas Development Institute. A background paper reviewing the links between income growth and nutrition was prepared by Edoardo Massset and Lawrence Haddad of the Institute of Development Studies. Summaries of Official Development Assistance for agricultural research and for food and nutrition security were provided by the Development Cooperation Directorate. An overview of the Policy Framework for Investment in Agriculture was provided by the Directorate for Financial and Enterprise Affairs. Statistical support was provided by Florence Bossard, with formatting and preparation by Anita Lari.

This document was declassified at the OECD Committee for Agriculture's Working Party on Agricultural Policies and Markets.

Table of contents

Tables

Figures

Boxes

Abbreviations

AFSI	Aquila Food Security Initiative
AMIS	Agricultural Market Information System
AR4D	Agricultural Research for Development
CAADP	Comprehensive African Agriculture Development Programme
CCTs	Conditional Cash Transfers
CRS	Creditor Reporting System
FAO	UN Food and Agriculture Organisation
FDI	Foreign Direct Investment
FNS	Food and Nutrition Security
GAEZ	Global Agro-Ecological Zones
GDPRD	Global Donor Platform on Rural Development
GHG	Greenhouse Gas emissions
GLOBALG.A.P.	Good Agricultural Practices
ICT	Information and Communications Technology
IOs	International Organisations
LDCs	Least Developed Countries
LIFDCs	Low-Income Food Deficit Countries
LMICs	Low Middle Income Countries
MDER	Minimum Daily Energy Requirement
MDGs	Millennium Development Goals
NFIDCs	Net Food Importing Countries
ODA	Official Development Assistance
OECD	Organisation for Economic Co-operation and Development
OLICs	Other Low Income Countries
PARM	Platform on Agricultural Risk Management
PFIA	Policy Framework for Investment in Agriculture
PPP	Public Private Partnerships
RIGA	Rural Income Generating Activities
RRF	Rapid Response Forum
SIDS	Small Island Developing States
SUN	Scale Up Nutrition initiative
SWAC	Sahel and West Africa Club
TFP	Total Factor Productivity
WHO	World Health Organisation
WTO	World Trade Organisation

Executive summary

The challenge of eliminating global hunger is more about raising the incomes of the poor than an issue of food prices.

Eliminating hunger and malnutrition, and achieving global food security more widely, is among the most intractable problems humanity faces. While many once poor countries are now developing rapidly, the world as a whole is unlikely to meet the First Millennium Development Goal target of halving, between 1990 and 2015, the proportion of the world's population who suffer from hunger. According to FAO figures, the total number of undernourished people in developing countries has fallen from just under a billion in 1990-92 to around 852 million in 2010-12. However, the pace of reduction has slowed and the absolute numbers remain stubbornly high.

The problem of hunger has been accentuated by high food prices. In low income countries, food consumption expenditures typically account for 50% or more of households' budgets. In lower middle income countries, such as China and India, the figure is about 40%. Farmers are affected by food prices as both buyers and sellers. Those with sufficient access to land and other resources may gain from higher prices, but a majority of the rural poor – including many farm households – are net buyers of food staples. Even short episodes of income loss can cause poor households to sell productive assets at low prices, leading to potential poverty traps.

There were legitimate fears that higher food prices could undermine the food security of millions. Recent data suggest that, while many households have faced undeniable hardship, the worst fears have not been realised. The chief reason is that robust income growth in many developing countries has been sufficient to outweigh the impacts of higher food prices. The global picture has been helped by strong growth in populous middle income countries, including China and India, where a large share of the world's undernourished live.

While price levels matter, they are not the fundamental problem. The persistence of global hunger – the chief manifestation of food insecurity – is a chronic problem that pre-dates the current period of higher food prices. Indeed there were as many hungry people in the world in the early 2000s, when international food prices were at all-time lows, as there are today.

The principal cause of food insecurity remains poverty and inadequate incomes. Globally, there is enough food available, although many people are too poor to afford it. Tighter world food markets, in which food is less available, make affordability an even bigger constraint. Broad-based income growth is therefore the key to lasting reductions in global hunger. Policies and investments that stimulate income growth are likely to reduce the need for short-term fixes that cope with consequences of low incomes but do not tackle the underlying causes.

Yet there is no need for anyone to be left unprotected. If people are too poor to afford food, then national governments can provide social safety nets and nutrition programmes. If governments do not have the required domestic resources, then funding gaps can be met by the international community. The UN Food and Agriculture Organisation (FAO) and the UN High-Level Task Force have proposed a twin track approach, consisting of an immediate response to the needs of vulnerable populations, combined with a commitment to longer-term strategies to address the chronic problem of undernourishment and to strengthen resilience to shocks. The Scale Up Nutrition (SUN) movement similarly proposes direct nutrition interventions, complemented by wider efforts to address the underlying causes of under-nutrition.

Agricultural development has a key role to play in ensuring food security.

Agricultural development has a key role to play in generating the incomes needed to ensure food security, especially in the poorest economies. Approximately two-thirds of the world's poor live in rural areas, where agriculture is the dominant sector. Most of the farming is done by smallholders, so raising their incomes is a priority. That can be achieved directly, by raising agricultural incomes, and indirectly by creating non-farm jobs and more diversified rural economies. Government strategies need to support both channels of development.

In a context of higher food prices, there are better opportunities for smallholder farmers to develop commercially viable operations than there have been for many years. Yet, the realisation of those opportunities by some smallholders will result in others moving out of agriculture into new, ultimately more remunerative, activities. Indeed, it is important to recognise that – as all OECD countries have experienced – the majority of future generations will have better opportunities outside agriculture than within it.

There is a need for increased investment in rural areas, which offer higher returns than agricultural subsidies.

A range of policies can strengthen the incomes of poor households, irrespective of whether they live in rural or urban areas, or work within or outside agriculture. The overall investment climate is central, and depends on fundamental factors such as peace and stability, sound macroeconomic management, good governance and developed institutions, clear property rights and adequate physical infrastructure. Improvements in education and primary healthcare can strengthen income growth, and improve nutrition directly.

In the case of farm households, these general policies have the benefit of raising their incomes, but do not deter them from taking advantage of non-farm opportunities as they emerge. At the same time, there is a strong case for increasing the share of public spending in support of agriculture and redressing urban bias in the allocation of resources. There are high returns to investments in agricultural research, technology transfer, and farm extension and advisory services. These investments help farmers directly; indirectly they benefit consumers by increasing overall food supply, thereby containing upward pressure on food prices and dampening the price volatility associated with tight markets.

In the case of low-income countries, it has also been suggested that – because of weak institutions and market failures – some market interventions may be warranted. For example, some price stabilisation has been proposed as a way of providing a more predictable investment climate and containing the impact of large international price swings. Similarly input subsidies for seed and fertiliser have been suggested as a way of redressing failings such as the under-development of infrastructure, missing markets for credit and inputs, and a lack of knowledge of the benefits of improved technologies. These arguments need to be balanced against multiple drawbacks. For example, price stabilisation thwarts the development of private risk management and can export instability onto world markets. Similarly, the provision of input subsidies can impede the development of functioning private markets. Moreover, such measures often become a target for special interests, imposing a severe drain on national budgets at the expense of essential public investments. If they are to be used, they should be time-bound with a clear exit strategy and they should not crowd out essential investments which tackle the market and institutional failures they are designed to offset.

Efforts to raise incomes need to be complemented by other policies to improve nutritional outcomes.

Countries where hunger is rife face many challenges, but the problems are not insurmountable. Income growth is necessary, but the composition of growth matters too. At the household level, more equal growth is likely to lead to faster improvements in the food security of the poorest. Inequalities in personal incomes are also often matched by inequalities in access to public services, such as education and primary health care. Universal provision of core public services would boost the potential of households to earn higher incomes. There are also direct benefits to nutrition from providing safe water and sanitation, and from specific initiatives to improve nutrition, such as improved awareness regarding adequate nutrition and child care practices, and targeted supplements in situations of acute micronutrient deficiencies.

As countries develop, the challenge of ensuring food security becomes progressively less a question of incomes and fiscal resources, and more one of modifying behaviour. Poor nutrition is a significant health issue in both developed and developed countries. Globally, there are more overweight people than there are underweight, while large numbers of middle income countries suffer from both problems, with significant proportions of the population underweight and overweight, and many individuals overweight yet poorly nourished. These issues may be more easily tackled via policies that raise consumer awareness and thereby change consumption habits, than through taxes and regulations. The scope for using food taxes to constrain demands is limited by the fact that most foodstuffs – unlike say tobacco – are good for health within limits.

The overarching supply-side challenge is to raise agricultural production sustainably while adapting to climate change. The OECD and FAO *Agricultural Outlook* suggests that structurally higher food prices are here for the coming decade. Strong demand and prices will provide farmers with the incentives needed to feed a wealthier world population that is expected to exceed 9 billion by 2050. But policymakers can further stimulate the food supply response and constrain demands that put upward pressure on food prices without leading to improved nutritional outcomes – for example by reducing waste throughout the food supply chain and encouraging consumers to adopt more balanced diets.

Sustainable agricultural productivity growth is central to ensuring that food will be available at prices people can afford.

There is more scope for raising agricultural productivity than there is for mobilising more land and water resources. While it is likely to become increasingly difficult to push yield frontiers at a constant percentage rate of growth (i.e. exponentially), there is great scope for developing countries to close the gap between actual and potential yields. The key to realising these gains is innovation in the wider sense, combining adapted technologies with improved farm management practices. There is evidence of high rates of return to research and development accompanied with extension, albeit with long time lags.

There is much less scope for increasing cultivated land area than there is for improving yields. Moreover, a large share of the world's agricultural production is based on the unsustainable exploitation of water resources. There is a need for policies to manage both land and water resources sustainably, for example by strengthening land tenure systems and introducing water charges or tradable water rights.

Climate change is expected to have a range of (mostly negative) effects on agricultural production. A range of investments – for example in research, irrigation and rural roads – can help improve resilience, but production will ultimately need to be located in areas where it is inherently sustainable. Accurate data and public information have a vital role in helping farmers to adapt.

There is important scope for sustainable intensification, and investments in infrastructure can help limit producer losses, which account for around one-third of all production in low income countries. Yet current production patterns may not always be compatible with sustainable resource use, implying trade-offs between sustainability and immediate food security outcomes. In many countries and regions, there is no effective pricing of natural resources, with the result that production is too intensive or occurs in areas where ultimately it should not. Pricing of resources could improve the sustainability of production but raise farmers' costs and, in some circumstances, put upward pressure on food prices. Likewise, agriculture is a major contributor to anthropogenic climate change, but taxing farmers' greenhouse gas emissions could lower their incomes and raise food prices. These trade-offs underscore the primary importance of income growth: only if incomes grow sufficiently can food security and sustainable resource be fully compatible. On the other hand, pricing of environmental services could raise some farmers' incomes.

Policies that subsidise or mandate the use of biofuels should be removed.

Another potential trade-off comes through the use of agricultural products as a source of renewable energy, with the diversion of land to biofuel production adding to upward pressure on food prices. There are huge uncertainties over the scale of impact that biofuels will have on overall land use. Technological developments in biofuels, the cost and availability of fossil fuels and the policy environment are hazardous to predict. The removal of policies that subsidise or mandate the production and consumption of biofuels that compete with food would imply that these technologies come on-stream when and where they make economic sense, and in the meantime do not jeopardise food security unnecessarily. Indeed, biofuels could provide significant economic opportunities for some developing country farmers.

Public investment, supported by development aid, can complement and attract private investment.

The connected challenges of raising agricultural and rural incomes, and boosting supply sustainably, call for large increases in agricultural investment. FAO estimates total investment needs in primary and downstream agriculture at over USD 80 billion per year over the next four decades, which is about 50% higher than current levels. Most of this investment will have to come from the private sector, but strategic public investments can help attract private investment – both foreign and domestic.

Many developing countries have a dearth of domestic resources, and their agricultural sectors have suffered from decades of under-investment. Rising levels of foreign investment, prompted by higher food prices, can help redress this neglect. However, there are legitimate concerns about the nature of some of these investments and who will benefit. Hence, it is important that governments provide appropriate framework conditions for investment in agriculture, and that there are commitments to responsible business conduct on the part of both investors and recipients. Development aid can be a catalyst, complementing the primary role of private sector investment.

Trade has an important role to play in ensuring global food security. Reforming countries may need to put in place parallel measures to maximise the benefits and reduce the costs.

Open markets have a pivotal role to play in raising production and incomes. Trade enables production to be located in areas where resources are used most efficiently and has an essential role in getting product from surplus to deficit areas. Trade also raises overall incomes through the benefits to exporters (in the form of higher prices than would be received in the absence of trade) and importers (through lower prices than would otherwise be paid), while contributing to faster economic growth and rising per capita incomes.

Nevertheless, there are legitimate concerns about potentially negative effects that may follow from greater trade openness, and how those effects should be managed. First, trade reform generates an immediate pattern of winners and losers. For protected farmers, trade liberalisation will lower the prices they receive and expose any lack of competitiveness. Equivalently, if exports are taxed then the removal of those taxes will increase consumer prices. Second, while domestic shocks may be more frequent and severe than international shocks, there have been episodic spikes in international prices that have been large enough to raise concerns about the immediate welfare of those who spend a large share of their budgets on food. Third, trade openness may lead some countries to import more of their food, and for some of them a spike in food prices that is not matched by increases in the prices of exports could lead to difficulty in meeting their food import bills. Fourth, there are concerns about the reliability of world markets. When food prices peaked in 2007-08, some countries failed to honour forward contracts and the widespread application of export restrictions to suppress domestic prices undermined some importers' confidence in world markets as a reliable source of food supplies. Fifth, on the nutrition side, there are possible downsides from increased trade, for example if the prices of energy rich but low nutrient food staples fall relative to the prices of more nutritious alternatives.

While acknowledging the legitimacy of these concerns, trade policy instruments are not the *optimal* tools for addressing them. In terms of the winners and losers created by trade reform, the needs of the latter are best addressed through a combination of adjustment assistance and social safety nets. Price support, and associated trade protection, tends to be inefficient at delivering support to farmers, and inequitably distributed. Moreover, among the poor (and hence food insecure) there are typically both buyers and sellers of food, so price instruments, and associated border measures, are particularly blunt instruments. In the case of potential exporters who should benefit from reform, there may be a need for complementary reforms and supply-side investments for those gains to be realised. Such measures may reinforce the gains even when there is existing capacity.

For mitigating the adverse impacts on incomes of international price volatility, targeted social programmes (including cash transfers) are a preferable option, while agricultural investments and the development of risk management tools can improve farmers' resilience to risk. Although price stabilisation (as opposed to price support) can limit the impact of adverse shocks on producers and consumers, it often proves to be fiscally unsustainable. As long as the programme endures, it can provide a more stable investment climate, but it thwarts the development of private risk management, and can export instability onto world markets.

Work on the macro implications of higher food prices suggests that self-sufficiency is likely to be an expensive way for food importing countries to limit their exposure to periodically higher food import bills. Hedging on international markets is an alternative option, while the international community has several financing mechanisms that could enable developing country governments to overcome rare but potentially severe surges, such as that experienced in 2007-08. Insofar as the prices of food items do not all move contemporaneously, countries can also limit their exposure to price risk by diversifying the commodity composition of both exports and imports.

The best way of coping with problems related to the unreliability of world markets is for countries to desist collectively from adopting beggar-thy-neighbour policies. These policies cause bilateral and regional trades to break down, and generate wider negative spill-overs when applied by countries with a larger presence on world food markets. Many of the responses to the 2007-08 food price spike were ineffective because of the collective impact of other countries applying similar measures. Countries can mitigate some of these risks by having a wider range of trading partners.

Finally, many countries face significant nutritional issues including not enough consumption and over-consumption (with often both occurring in the same countries) and unbalanced diets. Again the use of trade policy is a blunt instrument with which to try and modify consumption patterns. Public information and education are the first requirements for addressing these issues.

Trade will be essential in order for supply increases to be achieved sustainably. Trade enables production to locate in areas where natural resources, notably land and water, are relatively abundant, and where systems are more resilient to the effects of climate change. Looking ahead, the areas of the world with sustainable productive potential are not the same as the areas experiencing rapid population growth. Nor is there any one model of efficient farm structure. Global food security will need to be underpinned by a mix of small, medium and large farms, and by domestic as well as international markets.

Agricultural policy reforms by OECD countries would improve policy coherence...

OECD countries can accelerate the reform of policies that create negative international spill-overs. Agricultural protection remains high and many OECD countries continue to provide trade-distorting subsidies that constrain development opportunities for more competitive suppliers and can export instability onto world markets. While the prevalence of such policies has declined significantly, there is still much room for reform. In the current context of high agricultural prices, now is a good time to move rapidly towards alternative policy instruments that would contribute to sustainable productivity growth, underpinned by appropriate risk management and social protection policies. At the same time, OECD countries can avoid policies that contribute artificially to higher world food prices, most notably mandates for biofuel production.

...as would continued efforts to improve the functioning of world food markets

Getting world food markets to function more smoothly will also require wider efforts at the multilateral level. While WTO members may have come close to a new agreement on agriculture, conclusion of the Doha Round of trade negotiations remains elusive. G20 governments have sought to tackle two immediate dimensions of the food security question: how to combat price volatility and improve the functioning of world markets, and how to achieve sustainable agricultural productivity growth (and bridge the gap for smallholders). The OECD, along with other international organisations, has provided analytical support to those initiatives. Some of the recommendations are for specific policy changes, but equally important is the need to share knowledge on which policies work best and how to adapt policies to country-specific contexts.

The challenges to building global food security are increasingly understood, as are the ways in which more coherent and co-ordinated policies can accelerate progress. There are specific actions that OECD countries can take, and areas where action is needed at the global level. But national governments themselves have the responsibility for putting in place the conditions that will enable them to achieve food security for all their citizens.

Chapter 1

The challenge of global food security

This chapter describes the fundamental challenge of eliminating hunger and ensuring global food security. It assesses the scale of that challenge, identifies the basic conditions that need to be met, and sets out the key policy issues.

Ending hunger and malnutrition is among the greatest challenges humanity faces. Malnutrition is estimated to be the cause of 30% of infant deaths, the predominant factor behind the global burden of disease, and a major impediment to cognitive development, and to growth in labour productivity, wage earnings and overall incomes (Headey, 2013). With approximately 850 million people undernourished, the problem persists, despite technological advances in food production, unprecedented global wealth and rapid economic development in many parts of the world. It means solving the great paradox of hunger amid plenty. The world produces enough food for everyone, but many are too poor to afford it.

The highest profile commitment to tackle hunger, and focus for recent international efforts, has been through the Millennium Development Goals (MDGs). Goal 1 calls for the eradication of poverty and hunger. It includes specific targets of halving, between 1990 and 2015, the proportion of the world's people whose income is less than one dollar a day and the proportion of the world's people who suffer from hunger, measured via the prevalence of undernourishment and under-weight (i.e. an abnormally low weight-for-age ratio) among children under 5 years of age.[1] The combining of poverty and hunger targets within one goal implicitly recognises that the two are closely connected.

Progress on MDG1 has been uneven. According to FAO, the proportion of the population in developing countries that is undernourished has fallen significantly over the past two decades, from 23% in 1990-92 to 15% in 2010-12 (Figure 1.1). But the pace of decline has slowed and the world is not currently on target to meet the First Millennium Development Goal (MDG1) target of halving the proportion of undernourished people in developing countries between 1990 and 2015. Moreover, as a result of population growth, the total number of undernourished people in developing countries has fallen even more slowly, from just under a billion in 1990-1992 to around 852 million in 2010-12. This is far behind the more ambitious goal set at the 1996 World Food Summit, where countries pledged to eradicate hunger in all countries, with an immediate view to reducing the *number* of undernourished people to half the 1996 number by no later than 2015.

According to FAO data, 70% of the world's undernourished live in middle-income economies, mostly in Asia. Asian countries accounted for 65% of the world total in 2010-12, with the share of China and India alone above 40%, despite significant progress in China whose share of the total undernourished has dropped from 25% to 18% in ten years (Figure 1.2). On the other hand, the *prevalence* of undernourishment is highest in low income economies, at 30%. Africa is the most afflicted region, with 23% of people undernourished, compared with 14% in Asia, 8% in Latin America and the

Caribbean and 12.5% on average globally. The share of undernourished living in least developed countries (LDCs) has increased from one-fifth to one-third.

The WHO underweight data describe an even more disturbing situation than the undernourishment figures. Almost one out of five children under five was moderately or severely underweight in recent years (WHO, 2012) – a figure which has come down by 5 percentage points since the late 1980s.[2] As with the undernourishment information, the majority of the world's underweight children live in Asia. But in contrast with the undernourishment numbers, the prevalence is higher in Asia than in Africa, with a particularly high incidence in India, at over 40% (although that rate has come down from 60% in 1988-92). There are almost as many underweight children in India as in all of Africa.

Figure 1.1. Undernourishment in the developing world

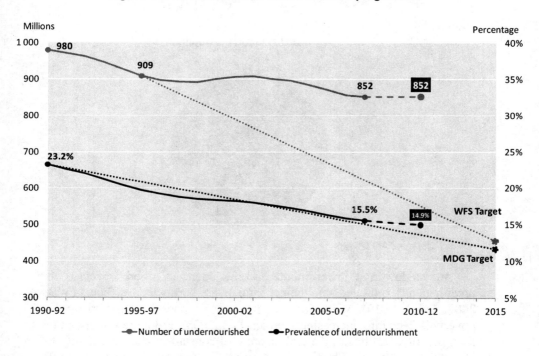

Note: WFS: World Food Summit; MDG: Millennium Development Goals.
Source: Adapted from FAO (2012a).

Figure 1.2. Global number of undernourished

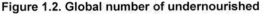

■ Developed countries   ■ Africa
▨ Latin America and the Caribbean   ▨ India
▨ China   ▨ Other Asia

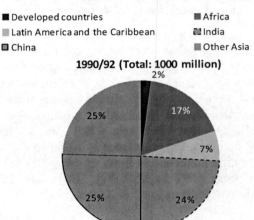

1990/92 (Total: 1000 million)

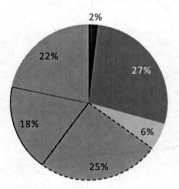

2010/12 (Total: 868 million)

Source: FAO (2012b).

Across low and lower-middle income countries, food insecurity is predominantly rural and smallholder farmers are particularly afflicted (WB, 2007; IFAD, 2010). While most of the world's poor live in rural areas, poverty is becoming increasingly urban (WB, 2008). From the standpoint of food security, that makes the real incomes of consumers and their ability to afford food an increasingly important issue. However, as income growth in middle income countries draws increasing numbers of poor households over the basic poverty threshold, the chronic problem of food insecurity is likely to be increasingly concentrated where growth still languishes – among the poorest farm households in the poorest parts of the world.

World Bank figures suggest that the target for reducing extreme poverty has already been met, with 22% of the developing world's population – or 1.29 billion people – living on USD 1.25 or less a day in 2008, compared with 43% in 1990 and 52% in 1981. Provisional estimates for 2010 indicate that the USD 1.25 a day poverty rate fell to less than half the 1990 rate, with developing countries generally managing to withstand food, fuel and financial crises. Moreover, the number of people in extreme poverty and the extreme poverty rate declined in every region of the developing world in 2005-08, for the first time since the World Bank started tracking extreme poverty (Chen and Ravallion, 2010).

The poverty data provide grounds for optimism, suggesting that major reductions in hunger and malnutrition are within reach. But they also suggest that raising incomes and reducing extreme poverty is not enough. Some countries have been more effective than others in translating income growth and poverty reduction into improved nutritional outcomes. Of the countries that have performed poorly in terms of nutritional outcomes, some have been marked by conflict, some have seen strong economic growth but the benefits of that growth have not reached the poorest, while in some countries essential complements to higher incomes – such as improved public sanitation and healthcare – have been missing.

Deficient incomes need not be an obstacle to adequate nutrition. If people are too poor to afford food, then there should in principle be ways of ensuring that they are properly nourished anyway. For example, national governments can provide social safety nets and nutrition programmes, while national funding gaps can be met by the international community. When world food prices spiked in 2007-08, a specially convened UN High-Level Task Force proposed a twin track approach, consisting of an immediate response to the needs of vulnerable populations, combined with a commitment to longer-term strategies to address the chronic problem of undernourishment and to strengthen resilience to shocks (UNHLTF, 2010). The Scale Up Nutrition (SUN) movement (SUN, 2012) sets out an agenda for effective backstopping, with direct nutrition interventions complementing wider efforts to address the underlying causes of under-nutrition.

The concept of food security sets out the overarching challenge. According to the FAO definition, agreed at the 1996 World Food Summit, food security exists when all people, at all times, have physical, social and economic access to sufficient, safe and nutritious food to meet their dietary needs and food preferences for an active and healthy life. Increased recognition of the importance of the nutritional dimension, for example at the 2009 World Summit on Food Security in Rome, has led many to prefer the term "food and nutrition security".

Food insecurity varies by time and degree. Chronic hunger typically affects very poor people who cannot afford to nourish themselves adequately. Hunger may be seasonal, with greater prevalence in the run-up to harvest when supplies are low and local prices high (Devereux, 2009). Populations can also be afflicted temporarily by food crises and emergencies. These attract more political and media attention than chronic food insecurity, but afflict smaller numbers of people (Wiggins and Slater, 2010). The important role of humanitarian relief in such circumstances is beyond the scope of this study.[3]

The emphasis of this study is on under-nutrition and the developing country dimensions of food insecurity. Strictly, food security covers a variety of nutritional situations, including over-nutrition and its consequences. Globally more than 1.4 billion adults were overweight in 2008, over a third of whom were obese; 65% of the world's population live in countries where overweight and obesity kill more people than underweight (WHO, 2012).[4] Large numbers of middle income countries suffer from both problems, with significant proportions of the population either underweight or overweight, and many individuals overweight yet poorly nourished. Overweight presents a major public health issue in developed, and increasingly developing, countries. These issues are addressed only to the extent that reduced over-consumption and re-balanced diets can reduce the demand for food, lower prices, and thereby improve the terms of access for poorer households. It is also important to acknowledge that people go hungry in developed OECD countries, and that poorer households often suffer from inadequate nutrition. Again, this is a significant public policy issue, but one that falls outside this study's focus on the functioning of the food and agriculture system.

The FAO's definition provides an organisational framework, suggesting that people will only be food secure when sufficient food is **available**, they have **access** to it, and it is well **utilised**. A fourth requirement is the **stability** of those three dimensions over time. The challenge is wide ranging, multi-faceted and linked to other huge agendas, including those of tackling world poverty, using scarce natural resources sustainably and managing and adapting to climate change.

Ensuring global food availability has not historically been a problem, and the real price of food has fallen dramatically since the end of the Second World War. But recent spikes in world food prices indicate that markets are getting structurally tighter and that an era of steadily declining real food prices has probably ended. Nevertheless, episodes of high food prices are not unprecedented, and recent price spikes are less severe than those experienced during the two world wars and in the 1970s, as data for the United States show (Figure 1.3).

Figure 1.3. Index of real US maize and wheat prices, 1908-2012

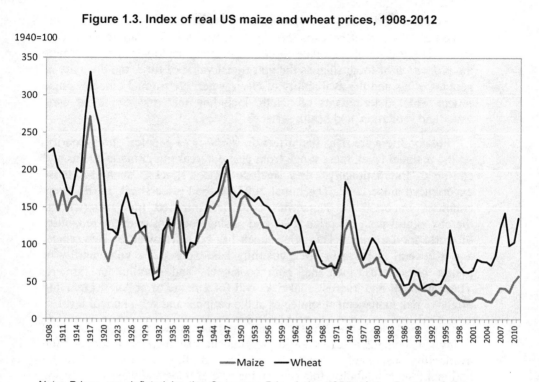

Note: Prices are deflated by the Consumer Price Index (CPI) of the Bureau of Labor Statistics (BLS).

Sources: OECD calculations based on USDA and BLS data.

Looking forward, the world's population has recently passed the 7 billion mark and, according to the UN's central projection, is expected to reach 9.3 billion by 2050. FAO estimates that, taking income growth into account, this will require a 60% increase in food production compared with 2005-07 (Alexandratos and Bruinsma, 2012). That translates into annual growth of 1.1% per year, which is lower than recent productivity growth (OECD and FAO, 2012). The challenge in terms of *availability*, however, relates to how the increase in food production is achieved: more food can be produced, but it must be done sustainably, taking into account constraints on natural resources and the effects of climate change.

The basic problem of food insecurity has been more on the food *access* side – poverty and deficient incomes – rather than on the availability side. The poor spend a significant share of their budgets on food and, until their incomes rise sufficiently, the cost of food remains an important determinant of their real incomes and access. The key to improved access is higher incomes.

Yet, among developing countries, there is a wide variation in nutritional outcomes that cannot be explained by differences in availability or access alone. These differences relate to complementary factors which determine the *utilisation* of food, such as the nutritional value of food, the diversity of peoples' diets and the availability of clean water. Nutritional outcomes also reflect wider determinants of health, including maternal and child care, water and sanitation, and health services.

Finally, there are risks that affect the *stability* of peoples' food security. At the national level, these range from pest outbreaks to climatic events and conflicts. Internationally, they include price shocks, such as those experienced since 2007. The initial 2007-08 food price shock raised many countries import bills. Numerous countries imposed trade restrictions, thereby aggravating the price spike and raising concerns over the reliability of world food markets. This price shock has been followed by two others, and a general increase in price volatility has exposed the vulnerability of poorer households, including both consumers and smallholder farmers. These national and international risks call for a range of policies, including effective risk management strategies at the national and sub-national level.

This study takes the FAO's definition and framework and applies it as follows. Chapter 2 assesses the basic challenge of increasing global food availability sustainably. Chapter 3 considers the links between global and national food availability, focusing on the role of trade and ways of ensuring the stability of national food supplies. Chapter 4 examines the determinants of peoples' access to food, as both producers and consumers. It considers the types of policies that can be effective in raising incomes and access and the role of risk management strategies in improving stability. Chapter 5 examines the utilisation dimension of food security, and the role of complementary policies in ensuring improved nutritional outcomes. Chapter 6 consolidates the main policy conclusions. These include recommendations for OECD countries, as well as for emerging and developing countries. They also identify areas where there is a need for global policy action to improve the functioning of world food markets.

It is important to note that a vast amount of research is underway and many organisations have produced important synthetic work on or related to the topic of global food security and its implications for the world's food and agriculture system. A major initiative was the UK Foresight project, which in 2011 produced a report entitled "The Future of Food and Farming: Challenges and Choices for Global Sustainability" (Foresight, 2011), the most thorough stock-taking so far of issues related to agriculture and food security. At the policy level, there are global efforts, notably the UN High-Level Task Force's Comprehensive Framework for Action and the Scale Up Nutrition (SUN) initiative, as well as regional ones such as the

Comprehensive African Agriculture Development Programme (CAADP). There is a wide range of work devoted specifically to this topic at FAO, whose State of Food Insecurity addresses these issues on a systematic and ongoing basis (FAO, 2012a). OECD has been engaged with other IOs in collaborative work for the G20 on issues pertaining to food security, with recent reports focusing on policy responses to price volatility and on productivity and innovation.

This study does not seek to summarise or challenge this important work. On the evidence side, the aim is to take stock of the current state of knowledge, identify areas where current OECD work is adding value to that knowledge and where future work can make an effective contribution. On the policy side, the objective is to produce information that can inform OECD countries' policies as well as multilateral initiatives, such as those pursued through the G20. It is also hoped that the material will contribute to the global debate on issues pertaining to food security.

Notes

1. The FAO's **undernourishment** indicator estimates the number of people who do not have access to enough food to meet its daily calorie requirement to live a healthy and active life. The estimation starts from the observation of food availability at the national level (converted to calorie equivalent), which is translated to the individual level on the basis of an estimated intra-national distribution of access to food. The quantity of calories to which each individual in the population is considered to have access is then contrasted with a minimum estimated energy requirement. People falling below this threshold are considered to be undernourished.

2. The WHO's index of the prevalence of "moderate **underweight**" is estimated as the proportion of children aged 0-5 years whose weight falls more than two standard deviations below the median of the reference population.

3. Readers are referred to the UN's Comprehensive Framework for Action, produced by the UN's High Level Task Force on the Global Food Security Crisis, which identifies actions to address the immediate needs of vulnerable populations and to build resilience (UN, 2010).

4. Overweight and obesity describe abnormal or excessive fat accumulation that may impair health. The WHO considers a person to be overweight or obese when her body mass index (a person's weight in kilograms divided by the square of his height in meters) is greater than or equal to 25 or 30, respectively.

References

Alexandratos, N. and J. Bruinsma (2012), "World agriculture towards 2030/50: The 2012 revision", *ESA Working Paper* No. 12-03, FAO, Rome.

Chen, S. and M. Ravallion (2010), "The developing world is poorer than we thought, but no less successful in the fight against poverty", *Quarterly Journal of Economics*, 125(4), pp. 1577-1625.

Devereux, S. (2009), "Seasonality and social protection in Africa", *Growth & Social Protection Working Paper* 07, Future Agricultures/Centre for Social Protection, Brighton.

FAO (2012a), *The State of Food Insecurity in the World*, FAO, Rome.

FAO (2012b), *Food Security Indicators*, Revised November, 27, FAO, Rome.

Foresight (2011), *The Future of Food and Farming: Challenges and Choices for Global Sustainability*, The Government Office for Science, London.

Headey, D. (2013), "The impact of the global food crisis on self-assessed food security", *Policy Research Working Paper* No. 6329, The World Bank, Washington, DC.

IFAD (2010), *Rural Poverty Report 2011*, International Fund for Agricultural Development, Rome.

OECD and FAO (2012), *OECD/FAO Agricultural Outlook 2012-2021*, OECD Publishing, Paris and FAO, Rome.

SUN (2012), "SUN movement: Revised road map", Secretariat of the Scaling Up Nutrition Movement.

WHO (2012), *World Health Statistics*, WHO, Geneva.

Wiggins, S. and R. Slater (2010), "Food security and nutrition: current and likely future issues", *Science Review* 27, Foresight Project on Global Food and Farming Futures, Government Office for Science, London.

World Bank (2007), *World Development Report 2008: Agriculture for Development*, The World Bank, Washington, DC.

Chapter 2

Ensuring global food availability

The chapter considers the ways in which governments can improve the availability of food sustainably. While food production will respond to the needs of a rising and more affluent world population, there are steps that governments can take to improve the availability of food, either by stimulating supply sustainably or by constraining demands that are detrimental to nutritional outcomes.

2.1. The challenge of ensuring global food availability

"The power of population is so superior to the power of the earth to produce subsistence for man, that premature death must in some shape or other visit the human race." – Malthus T.R. 1798. An essay on the principle of population. Chapter IX, p. 72.

Despite Malthus's gloomy prediction, the overall availability of food has not historically posed a problem for global food security. While demand has increased as a result of population growth and rising incomes, production has kept pace and there has been no sustained period over which population growth has outstripped supply. Over the past 50 years, the amount of food available per person has increased by 20% (Figure 2.1). Availability has more commonly been an issue at the national level, but even then it has not been the dominant cause of famine. The broad evidence confirms Sen's overall assessment that food access matters most: "Starvation is the characteristic of some people not having enough food to eat. It is not the characteristic of there being not enough food to eat" (Sen, 1980).

Figure 2.1. Global food production and population growth

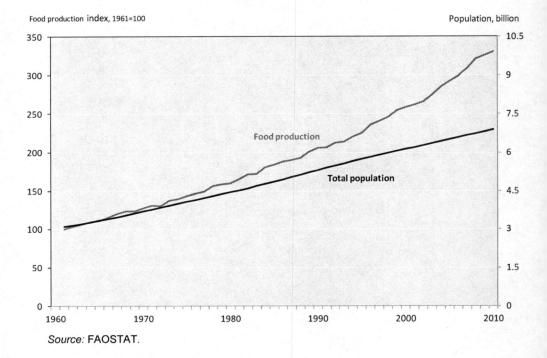

Source: FAOSTAT.

The key issue with respect to global food availability is the prospect of tighter world food markets, with demand increases, deriving principally from income and population growth, outpacing expected supply gains coming from productivity improvements and increased mobilisation of land, water and other resources. Tighter world markets imply higher, and possibly more volatile, food prices. Thus, the problem of availability becomes one of access for those who can no longer afford food.

Increases in food availability, which contain or reverse upward pressure on food prices from population and income growth, can be achieved by stimulating supply, or by restraining demands that do not correspond to improved "utilisation" of food. The main channels through which governments can improve global food availability are noted in Table 2.1.

Table 2.1. Ways of increasing global food availability

Increasing food supply	Limiting food demand
Improved agricultural productivity (more efficient use of inputs, such as labour, land and water)	Modified tastes and preferences (including less meat consumption, reduced over-consumption)
Expansion of land area	Reduced consumer waste
Reduced supply chain (especially post-harvest) losses	
Climate change adaptation	
Less diversion of crops to non-food uses (e.g. biofuels)	

Conventional agricultural policies, such as price and farm income support and credit subsidies, also have effects on the supply side, while food taxes and consumer subsidies affect demand. Trade also has an important role to play in increasing aggregate food availability, with open trade enabling food production to locate to areas where it can be undertaken relatively efficiently and providing a mechanism through which food can be allocated from surplus to deficit countries and regions. The role of trade in contributing to global food availability is taken up in Chapter 3. Yet the objective of increasing food availability cannot be viewed in isolation. Those increases need to be generated efficiently (i.e. policies need to be cost-effective) and sustainably.

In terms of efficiency, the basic questions are: First, what changes to the supply and demand factors listed above are likely to occur and to what extent can they be influenced by policies? Second, how much would it cost to effect those changes? Answering such questions should enable governments to prioritise.

In terms of sustainability, there may be complementarities. The broad challenge of "sustainable intensification" is to exploit those complementarities, i.e. to increase agricultural productivity growth without imposing greater strain on natural resources, in a context of growing competition between agriculture and other uses for finite land and water resources, and uncertainties associated with climate change and other environmental problems (FAO, OECD et al. [for G20], 2012). It will require adopting technologies and farm management practices that reduce GHG emissions, sequester carbon, adapt to climate change and provide environmental co-benefits. Recent OECD work explores how the cultural and social changes, effected for example via education and the provision of information, can facilitate adaptation to, and mitigation of, climate change by farmers (OECD, 2011a).

However, it is important to note that there may also be unavoidable trade-offs. In particular, farmers may be located in areas where production is not inherently sustainable. Relatedly, there may be cases where production is occurring without effective pricing of natural assets, and without taxing negative externalities. For example, case studies commissioned by the International Sustainability Unit provide several examples where the market price of food is lower than the true costs of its production. In particular many production practices impose negative externalities and erode natural capital, depriving future generations of natural resources (ISU, 2011). The sources of loss include greenhouse gas emissions, air and water pollution, soil degradation, water depletion and losses in biodiversity. A common problem is unsustainable irrigation practices. For example, IFPRI modelling work suggests that over-exploitation of water resources in Punjab and Haryana (partly attributable to free electricity, which leads to excess use of electric pumps) may lead to a decline in wheat production of around 15% by 2020. The net present value of this loss is estimated at about USD 1.2 billion (ISU, 2011).

These examples indicate that the pursuit of environmental sustainability may not always be consistent with raising food production. If policy makers are reluctant to tax negative externalities or to price natural capital because of the implications for a particular constituency's livelihoods, then it is important that any trade-offs are at least made clear. The costs of not pricing resources for sustainable use can then be viewed as an implicit subsidy to farmers (and indirectly to consumers), necessary to guarantee their short term food security. Over time, it should be possible to phase that subsidy out as income growth outweighs the burden of higher costs and food prices, and as farmers are encouraged to transition to more sustainable farm practices or to alternative livelihoods that can generate higher incomes.

The benefits from changes to the factors in Table 2.1 would go beyond increased food availability and lower prices.[1] Most of the supply side changes would also lead to higher farm incomes; while on the demand side, reduced over-consumption and a shift to more balanced diets in some countries would lead to improved health. Likewise, reducing waste on either the producer or consumer side would reduce resource pressures. These additional impacts are taken up in later sections.

In terms of prioritising among policies, it is helpful to take stock of what world food availability would look like under a plausible "business as usual" scenario, then consider the scope for policymakers to shift the basic supply and demand determinants and the implications of doing so. Section 2.2 presents the main characteristics of the outlook for world food and agricultural markets over the next ten years, drawing on the OECD and FAO *Agricultural Outlook* and the underlying Aglink-Cosimo model. It also distils the main findings from a range of modelling efforts which address expected changes in food availability over the coming decades – out to 2050 and in some cases beyond. Following that, Section 2.3 looks at the main supply shifters, and considers the nature of the link to food security outcomes and potential policy responses. Section 2.4 does the same for the demand shifters.

2.2. Outlook for world food availability

OECD works with FAO to produce an annual OECD–FAO *Agricultural Outlook*, which provides projections for world agricultural markets over the medium term (i.e. with a ten-year horizon) on the basis of a jointly maintained model (Aglink-Cosimo). At the same time, OECD participates in the Agricultural Model Inter-comparison and Improvement Project (AgMIP), which forms the basis for longer term scenario analysis.[2]

Aglink-Cosimo is a global partial equilibrium model of world agricultural markets which provides the baseline projections for agricultural commodity supply, demand, trade, and prices reported in the annual *Agricultural Outlook*. The strength of Aglink-Cosimo comes from its extensive country and commodity coverage, with 39 individual countries, 19 regions and 17 products or groups of products for which market clearing prices are specified (covering wheat, rice, coarse grains, oilseeds, oilseed meal, vegetable oil, sugar, beef, pork, poultry, eggs, and milk and key milk products; and in the most recent version, ethanol and biodiesel). The current *Outlook* is summarised in Box 2.1.

AgMIP seeks to clarify the links between market developments, climate change and food security. AgMIP participants include multiple groups working with crop models, agricultural economics models and world agricultural trade models. Within the overall AgMIP framework, the global economic models take inputs from crop and more detailed regional economic models. They then seek to harmonise core assumptions across the various models in order to make comparisons across different modelling approaches meaningful. These assumptions produce a reference scenario, and form the basis for an exploration of alternative scenarios. The models take a longer term perspective, exploring the implications of different scenarios through to 2050. The inter-comparison work of the AgMIP global economic models group contains both partial equilibrium and general equilibrium models. The AgMIP models typically have less commodity detail than Aglink-Cosimo but a more explicit treatment of factor markets, and are better placed to handle issues such as land and water constraints and climate change effects. Beyond AgMIP, a range of other modelling efforts are also underway, exploring the long term implications of alternative policy scenarios (e.g. Alexandratos and Bruinsma, 2012; Hertel, 2010; Paillard et al. 2011).

The different models shed light on different elements of the food availability issue, and are used to analyse a wide range of possible future developments and their driving factors. It is not possible to summarise all these modelling efforts or discuss their strengths and weaknesses. Instead, this section distils what they have to say about the core forces driving world food availability over the coming decades and the scope for raising food availability via each channel.

Box 2.1. Summary of the OECD and FAO Agricultural Outlook

Under the baseline assumptions, agricultural commodity prices will remain high throughout the next decade. High prices are driven by the eventual strengthening of global economic growth and stronger demand for agricultural products, along with growing biofuel demand and slowing production growth. Higher oil prices (foreseen to increase from USD 111 per barrel to USD 142 per barrel by 2021, an average annual growth rate of 2.9%) raise the costs of fertiliser and chemicals, and contribute to slowing productivity growth. Resource pressures, which include limited land and water availability, also imply that area expansion slows. The combined result is slower production growth and less accumulation of stocks. Aglink-Cosimo projects that prices of all commodities covered in the *Outlook* will be higher in nominal terms in 2012-21 than in the previous decade. When expressed in real terms (i.e. adjusted for inflation) all commodity prices apart from wheat and rice will be higher than their average in the previous decade 2002-11. When comparing the *Outlook* period with the averages of 2009-11 all crops show prices below the peak reached in 2011.

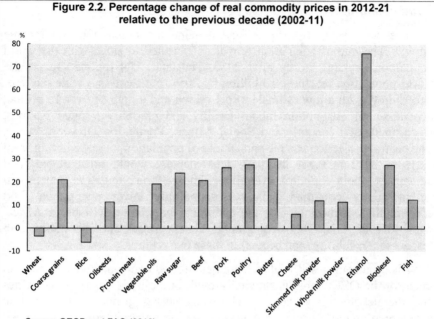

Figure 2.2. Percentage change of real commodity prices in 2012-21 relative to the previous decade (2002-11)

Source: OECD and FAO (2012).

The projected real price increases over the coming decade are higher for livestock products than for crops. The *Outlook* suggests that one reason is that many of these products did not experience a surge in 2007/08 as occurred for cereals and oilseeds. The smaller rise in feed costs relative to projected meat prices will improve margins in the livestock sector, which together with increased demand, will provide incentives to increase livestock inventories over the *Outlook* period. Rising per capita consumption of fish products will push up fish prices from both capture and aquaculture, the latter expected to increase more rapidly due to higher input costs. Despite strong meat prices, meat imports of developing countries are expected to increase, driven by population and income growth, in conjunction with high income elasticities of demand (OECD and FAO, 2012).

After the turbulence in recent years, the large rebound in supplies of major crops in response to high prices has helped to restore market balances. The projected higher prices are expected to encourage producers of crop and livestock products to increase area harvested and animal inventories; and to achieve higher productivity through further investments (e.g. use of improved seed varieties, inputs and high quality feedstuffs, adoption of productivity enhancing technologies in the face of rising energy prices). With increased commodity supply expectations and rising stocks, the risks of high price volatility are expected to abate in the near term. However, the *Outlook* notes that any unforeseen production shortfalls or trade restricting measures in major producing and trading countries could quickly provoke price rebounds and higher volatility (OECD and FAO, 2012).

Food demand

The core drivers of rising food demand are population and income growth. The rate of population growth is expected to slow, with the world population peaking shortly after 2050. The latest UN figures suggest the world population reaching 9.3 billion by 2050, but there is a wide range of uncertainty, with a low estimate of 6.1 billion and a high of over 15 billion, depending on assumptions about fertility and mortality (United Nations, Department of Economic and Social Affairs, Population Division, 2011). The central projection has the annual rate of population increase falling from 1.2% in 2012 to 0.3% in 2050. This increase would raise global food demand by about one-third, even though most of the population growth will be in poorer countries with correspondingly lower per capita food consumption. Almost half of the additional population will be in Africa, with 40% in Asia. This demographic change raises specific issues with respect to availability (and access) in these two regions.

Rising incomes will lead to increases in food demand. Weak demand in much of the OECD area is slowing growth in the large emerging countries and the developing world, but ultimately strong growth is expected in developing countries, with incomes converging towards those in developed OECD countries (OECD and FAO, 2012). Hertel et al. assume a global per capita income growth rate of 2.25% per year (Hertel et al., 2012).

Higher incomes will also change the composition of food demand, with more demand for livestock products in particular, but also for fruit and vegetables, as well as for sugar and vegetable oils. Tweeten and Thompson (2008) calculate that the combined impact of growing incomes and changing diets has been stable growth in per capita demand for food and fibre of around 0.27% per annum (measured over the period 1961-2000). Over the 45 years to 2050 this adds just 13% to aggregate food demand. However, the FAO (FAO, 2012a) suggests a per capita increase of around 30% over the same period, while Tilman et al. (2011) provide an estimate of 60%, showing the range of uncertainty.[3]

Taking population and income growth together, FAO estimate that, by 2050, global agricultural production will need to increase by 60% overall compared with 2005-07, and by 77% in developing countries, to meet rising demand, with per capita calorie consumption reaching 3 070 per day – considerably higher than a healthy level (FAO, 2012a). This implies an additional annual consumption of 940 million tonnes of cereals and 200 million tonnes of meat by 2050.

Supply response

Demand for feedstocks for biofuels has been an important factor behind renewed growth in cereal demand (Figure 2.3). These changes have been driven by a combination of high oil prices, changes in technical regulations, government mandates and other public policies. But if oil prices increase at the rates projected in the IEA's *World Energy Outlook*, Hertel et al. (2012) argue that in the long run biofuels will be competitive without subsidies and greenhouse gas emissions targets.

There is growing evidence that climate change has had and will have negative effects on agriculture, especially in developing countries.[4] In the near term, climate variability and extreme weather shocks are projected to increase (FAO, 2011). However there is a high level of uncertainty regarding the magnitude and direction of different effects.

The indirect effects of increased GHG emissions will differ widely across different regions. For example, high latitude areas could see an increase in their agricultural potential because of warmer temperatures, while regions near the equator will experience more frequent and severe droughts, excessive rainfall, and floods which can destroy and put food production at risk. At the same time, the capacities of economies to adjust to the effects of climate change depend on the socio-economic and technological conditions and political processes (Foresight, 2011). Moreover, increased GHG emissions are expected to have a direct effect on agricultural production through the positive response of plant growth to higher carbon dioxide concentrations; but increases in temperature above a given level lead to a decrease in efficiency of photosynthesis and an increase in respiration, hence a decline in productivity (FAO, 2011).

Modelling all these aspects is highly complex, and estimates of the magnitude of impacts vary according to models and scenarios. Tubiello and Fischer (2007) found impacts on world cereal production ranging between -18% and +18% for different regions by 2080. On the other hand, Fischer concludes that the impacts by 2050 on world cereal production will be modest, with declines by between 0.2% and 0.8% overall, and by between 0.2% and 4.2% in developing countries (Fischer, 2009). Estimates for crops vary depending on whether they are rainfed or irrigated. For example, according to IFPRI simulation results (Nelson et al., 2010), global yields would fall by about 7% in the case of irrigated maize, and by 12% for rainfed maize, between 2000 and 2050 in the absence of mitigation or adaptation policies. For rice the global yield reductions would be about 12% for irrigated rice but almost zero for rainfed rice. These global averages mask large disparities between developed and developing countries:

reductions in maize yields range between about 12% (irrigated) and 30 % (rainfed) in developed countries compared with 3% (irrigated) and 0.5% (rainfed) in low-income developing countries.

Figure 2.3. Growth in global cereal demand

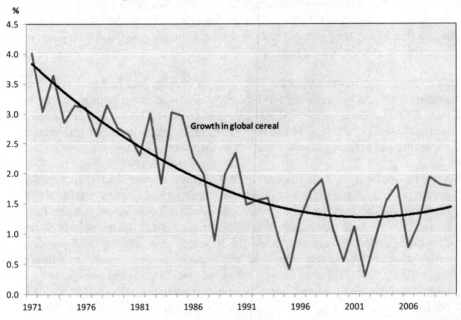

Note: The annual growth rates are calculated as ten-year averages for the ten years up to and including the year for which the annual growth rate is shown. The trend is fitted as an exponential curve to the annual growth rates.

Source: FAOSTAT.

The FAO's projections from 2006 did not take explicit account of the emergence of biofuel demand, or factor in the impacts of climate change. Fischer (2009) found strong effects from biofuels, with these adding between 4% and 35% to cereal prices, depending on the scenario. The price impacts are sensitive to the share that first-generation biofuels are mandated to contribute to total transport fuel consumption. On the other hand, climate change effects did not much alter the projected level of world prices in 2050, with changes of between -2% (with CO_2 fertilisation) and +5% (without CO_2 fertilisation).

Over the medium term, the Aglink-Cosimo baseline projects that real agricultural prices will be higher in 2012-20 than in 2002-11, with recent price spikes a harbinger of structural change in world food markets. But over the longer term, there are huge uncertainties about each of the core drivers of supply and demand, which make forecasting hazardous. There are wide divergences between upper and lower bound estimates on population and income growth. As noted in the sections below, there is wide scope for improving productivity, changing dietary patterns, and reducing waste on both the producer and consumer sides. Outside the agriculture system, the availability of new energy sources, such as shale gas, could have profound implications for food markets. In terms of prices, the possibilities include real prices going either down or up. But there is clearly a risk of much higher prices. IFPRI's pessimistic scenario, using their IMPACT model, suggest that by 2050 rice prices could be 78% higher than in 2010, wheat prices 59% higher, and maize prices 106% higher. With perfect climate change mitigation (but with the same pessimism on other factors) those increases would drop to 20% for rice, 24% for what and 34% for maize (Nelson, et al., 2010).

On the positive side, the world food system is flexible and contains important built in stabilisers (Hertel, 2010). A large increase in demand, which would cause prices to rise, will not only bring more land into production (the extensive margin), it will lead to increased yields on land (the intensive margin). Higher prices will also curb demand. Hertel argues that many of the models currently in use underestimate the importance of these built-in stabilisers by using relatively lower short-term elasticities rather than more appropriate higher long-term elasticities. Moreover, it is useful to bear in mind that projected population growth and consumption pattern changes suggest a 60% increase in food production between 2005-07 and 2050 (Alexandratos and Bruinsma, 2012). That translates into annual growth of 1.1% per year, which, as described in the next section, is lower than recent productivity growth. To summarise, increasing food demand imposes a daunting supply-side challenge, but one to which the evidence suggests the world's agricultural system is capable of responding.

Price volatility

Beyond the level of prices, a range of factors may contribute to increased price volatility. One is the prospect of a closer link between food prices and oil prices. Oil prices affect agricultural input prices directly and indirectly (through the price of fuel and fertiliser). In addition, depending on the relative prices of agricultural crops and oil, biofuel production may become profitable (without government support) in some OECD countries. At the same time, biofuel mandates can contribute to food price volatility by creating a supply for non-food use that is unresponsive to price. Other factors that could contribute to increased price volatility include lower stocks-to-use ratios than in the past, climate change impacts, the shift of production to new areas with more uncertain yields, and growing pressure on scarce resources (FAO, OECD et al. [for G20], 2011).

There is plenty that can be done to mitigate price volatility. Deeper integration of global and regional markets, better defined safeguard mechanisms and improvements in the competitive environment will bring increased trade volume and more suppliers and buyers to markets that are currently shallow. Local or regional supply shocks could more easily be absorbed, leading to lower volatility on domestic and international markets, and food could more easily flow from surplus areas to rapidly urbanising food-importing countries. Successful conclusion to the WTO Doha Development Agenda negotiations would be an important step, along with complementary policies that improve supply capacity and ensure the benefits of open and competitive markets are widely spread (FAO, OECD et al. [for G20], 2012). The extent to which financial speculation might be a determinant of agricultural price volatility is subject to disagreement, but well functioning futures markets for agricultural commodities, could play a significant role in reducing or smoothing price fluctuations – indeed, this is one of the primary functions of commodity futures markets.

2.3. Easing supply constraints

Achieving sustainable agricultural productivity growth

Increased productivity offers more scope for increasing food production than mobilising more resources. Fuglie (2012) estimates that increases in total factor productivity (TFP), broadly defined as total outputs over total inputs, accounted for three-quarters of global output growth in 2001-09. This compares with less than 7% in 1961-70 when output growth was mainly driven by increases in land and other input use. In OECD exporting countries, growth in output is almost all due to TFP growth, not to higher input use. According to World Bank and FAO estimates, yield improvements of the three most important cereals (rice, wheat and maize) rather than area expansion have been the basis for production increases over the last 50 years (World Bank, 2012a). Similarly, Bruinsma (2011) decomposes the historical growth in world crop production over the 1961-2005 period and finds that 77% of this growth came through yield growth and 9% from increased cropping intensity, with just 14% due to expansion in arable land area, although these components differed by crop.

There is some lack of consensus on whether agricultural productivity growth has been increasing or decreasing. According to USDA-ERS estimates, total factor productivity (TFP) growth in the past two decades has exceeded 2% per year in both developed and developing countries, comfortably outpacing world population growth, which is currently running at around 1.1% per year (Fuglie, 2012). Output growth rates have fallen, but input growth rates have fallen by even more (Figure 2.4.). In developed countries, resources were being withdrawn from agriculture at an increasing rate. TFP continued to rise but the rate of growth in 2000-07 was under 0.9% per year, the slowest of any decade since 1961. In developing regions, input growth slowed but was still positive, while productivity growth accelerated in the 1980s and following decades. Two large developing countries in particular, China and Brazil, have sustained exceptionally high TFP growth rates since the 1980s. Performance has been less encouraging in some countries and sub-regions. In particular, sub-Saharan Africa as a whole lags behind, with TFP growth rates of less than one per cent. Also, Asia's performance has been modest if one nets out the strong performance of China (Fuglie, 2012). In the 1990s, factor inputs contracted sharply in the countries of the former Soviet Union and output fell significantly. However, by 2000, agricultural resources had stabilised and growth resumed, led entirely by productivity gains in the sector.

Figure 2.4. Trends in total factor productivity growth for world agriculture

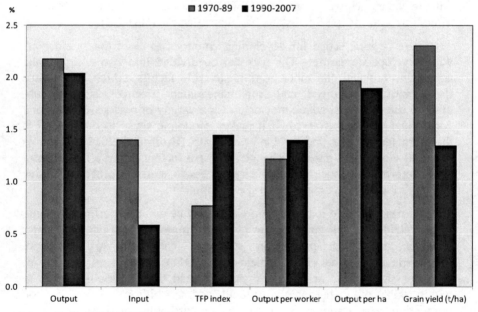

Source: Fuglie (2012) from FAOSTAT.

On the other hand, crop yields, used as an indicator of land productivity, show declining average global rates of growth for most of the major cereals (FAO, OECD et al. [for G20], 2012 and Alston, 2010). In many countries and regions yields are well below both their genetic potential and their potential in an economic sense, i.e. in terms of exploiting differences between the benefits and cost of attaining a given increase in output. Crop yields in Sub-Saharan Africa and South Asia remain, in most cases, much lower than in other regions, with cereal yields in Central and West Africa of 1-2 tonnes per hectare, contrasting with average yields of 7 MT/ha for wheat and 9 MT/ha for maize in Western European countries. There is also a wide divergence in rice yields across Asia, with yields of less than 4 MT/ha in Southern and Central Asia, and only 2 MT/ha in India, contrasting with yields of over 6 MT/ha in East and West Asia. In aggregate terms, developing countries are closing yield gaps with the most productive OECD countries, but this convergence does not extend to many of the world's poorest economies.

At the global level, a greater share of future productivity improvements is expected to come from improvements in technical efficiency (moving closer to the boundary of the production possibility frontier) rather than through technological change (moving the frontier forward), with the latter slowed by diminishing returns in plant and livestock breeding, and by climate change. However, a recent OECD study suggests that biotechnology offers important scope for sustainable intensification (Box 2.2).

There is great scope for developing countries to close the "yield gap" with developed countries. The gap can be divided into two components: agro-environmental and other non-transferrable factors, which create gaps that cannot be narrowed, and crop management practices, such as sub-optimal use of inputs, which may occur for a variety of reasons. The second component can be narrowed, if it makes economic sense to do so, and is therefore termed the "bridgeable" yield gap (Bruinsma, 2011). There is scope to close yield gaps by changes in these factors: more efficient farm sizes, improved management capacity, access to markets, other legislative and institutional factors, and better use of inputs.

The best places to improve crop yields may be on underperforming land, where yields are currently below average. Improved nutrient and water supplies and other production strategies can lead to significant improvements in crop yields. Foley et al. (2011), in a recent analysis of 16 major staple food and feed crops, estimated that increasing yields to within 95% of their potential would add 2.3 billion tonnes to crop production, representing a 58% increase over current production.

Mueller et al. (2012) find that closing yield gaps to 100% of attainable yields could increase worldwide crop production by 45% to70% for most major crops (with 64%, 71% and 47% increases for maize, wheat and rice, respectively). Eastern Europe and Sub-Saharan Africa show considerable "low-hanging" intensification opportunities for major cereals; these areas could have large production gains if yields were increased to only 50% of attainable yields. Looking forward, the OECD and FAO *Agricultural Outlook* anticipates lower yield growth over the coming decade due to increased pressure on natural resources. At the same time, Ludena et al. (2007) project that TFP growth will accelerate over the coming decades. The latter assumes faster land productivity growth in the livestock sector, and more rapid improvements in technical efficiency (i.e. factors being combined more efficiently).

The key to wider total factor productivity improvements is innovation, which the *Oslo Manual* defines as "the introduction of new or significantly improved goods or services, or the use of new inputs, processes, organisational or marketing methods" (OECD and Eurostat, 2005). The concept of an "innovation system" takes account of the interactions of individuals and organisations processing different types of knowledge within particular social, political, policy, economic and institutional constraints. Innovation systems are increasingly linked to the adoption of more sustainable, as well as more productive, practices, such as no-till farming, the development of insect resistant crops, more efficient irrigation and better water management systems.

Shortcomings in crop management practices may be overcome by agricultural education and wider investments in human capital, together with more effective use of inputs. Wider constraints to yield growth include a lack of access to output and input markets, due to trade barriers, monopoly power or weak infrastructure. Institutional and legislative factors may also be important, for example in facilitating or thwarting the emergence of efficient farm structures (including efficient farm sizes).

Investment in the agricultural sector is strongly correlated with TFP performance. Evenson and Fuglie (2012) found TFP performance in developing-country agriculture to be specifically correlated with national investments in agricultural research and technological improvement, and the country's ability to develop and extend improved agricultural technology to farmers ("technology capital"). Countries that had failed to establish adequate agricultural research and extension institutions and extend basic education to rural areas were stuck in low-productive agriculture and behind the rest of the world.

Box 2.2. The use of biotechnology in agriculture

Biotechnology offers technological solutions to the challenge of increasing agricultural production subject to finite resources (notably land and water) that are likely to be further constrained by climate change. It includes not only genetic modification (GM) but also intragenics, gene shuffling and marker assisted selection. These techniques can increase the supply and environmental sustainability of food, feed and fibre production, improve the nutritional content of food staples and help to maintain biodiversity (OECD, 2009). In conserving scarce natural resources they are a potentially important complement to improved agronomic practices (Rosegrant et al., 2012).

OECD work estimates that by 2030 approximately 50% of agricultural output could come from plant varieties developed using one or more types of biotechnology – even without accounting for use in biofuels or as biomass for industrial feedstock. Many challenges of using crops for biofuel and other non-food uses could be addressed through biotechnology, allowing crops to be adapted for growth in different environments, raising productivity or increasing the efficiency of processing (Rosegrant et al., 2012). The OECD study notes that approximately 75% of the future economic contribution of biotechnology and large environmental benefits are likely to come from agriculture and industry, yet over 80% of research investments in biotechnology by the private and public sectors go to health applications.

The OECD report recommends that member countries: (i) boost research in agricultural and industrial biotechnologies by increasing public research investment, reducing regulatory burdens and encouraging private-public partnerships; and (ii) encourage the use of biotechnology to address global environmental issues (e.g. climate change) by supporting international agreements to create and sustain markets for environmentally sustainable biotechnology products.

Gene modification technology has created economies of scope and scale that have driven rapid corporate concentration. However, there is greater scope for the development of collaborative networks, and small dedicated biotechnology firms – as are common in the health sector. On the production side, the use of biotechnology can disrupt existing business models, implying a need to manage structural change away from existing production methods.

Some of the challenges for agriculture are social and institutional, including public opposition. Social attitudes to biotechnology can influence market opportunities, driving firms to alter the type of biotechnology used. Public opinion can also change if there is effective regulation and biotechnology products are seen to provide benefits for consumers and the environment. The OECD study stresses the importance of creating an active and sustained dialogue with society and industry on the socio-economic and ethical implications and requirements of biotechnologies.

Source: OECD (2009).

Box 2.3. Agricultural research for development (AR4D)

The 2009 L'Aquila statement on global food security called for strengthened investment in access to education research, science and technology, as analyses of the impact of AR4D show that such investments have a very high rate of return.

Applying a narrow definition of AR4D (i.e. only Creditor Reporting System category 31182 – agricultural research), total Official Development Assistance (ODA) expenditures averaged USD 471 million per annum over 2009-10. About 20% of the overall total came from the multilateral sector, while France is by far the major bilateral donor, accounting for just under half of the bilateral total.

Actual support for AR4D is, however, expected to be much higher as some DAC donors may be reporting ODA for AR4D under other sector codes. Therefore, taking a broader definition of AR4D that covers the wider "agricultural education/research/services grouping," total ODA expenditures averaged USD 1.3 billion per annum in 2009-10, representing 11% of total ODA for Food and Nutrition Security (FNS). France is again the main donor and its ODA is dedicated primarily to agricultural research. Other important donors such as Canada, Denmark, Japan and the United States focus much more of their ODA on AR4D on agricultural, livestock and financial services (Figure 2.5).

AR4D can make an important contribution to FNS, but only relatively small amounts of aid presently support these activities. The Aquila Food Security Initiative (AFSI) group has therefore decided to monitor progress on the commitment to increase investment in this area and to align it better with partner countries' identified priorities.

Figure 2.5. Bilateral ODA for AR4D: 2009-10 average (million, USD)

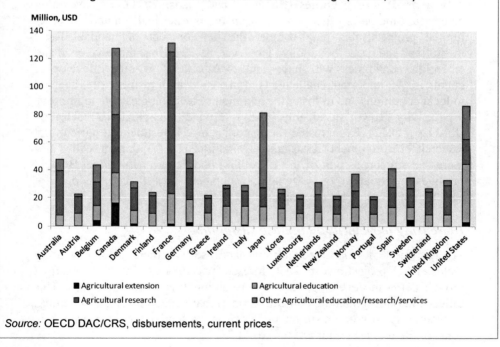

Source: OECD DAC/CRS, disbursements, current prices.

Moreover, there is specific evidence of high returns to spending on agricultural R&D, implying under-spending, especially in developing countries. Annual internal rates of return of investments on agricultural R&D estimated in the literature range between 20% and 80% (Alston, 2010). In developing countries, the dollar-for-dollar impact of R&D investments on the value of agricultural production is generally within the range of 6% to 12% (Fan et al., 2008, Fan and Zhang, 2008; FAO, 2012b). Those countries which have invested heavily in R&D while simultaneously investing in extension have had the strongest productivity growth (Fuglie, 2012). Given long time lags, it is likely that the high returns to R&D are also associated with the progressive adoption of innovation in the wider sense.

Government expenditures on agricultural R&D in developing countries are generally lower as a percentage of agricultural GDP than in OECD countries, but there is a wide diversity across countries in terms of percentage shares and growth rates (OECD, 2011b). China accounted for about two-thirds of total public agricultural R&D spending in low- and middle-income countries in 2002. China's agricultural research spending accelerated rapidly during the 1981–2007 period, especially after the turn of the millennium (FAO, 2012b). In Sub-Saharan Africa, after a decade of stagnation in the 1990s, investment in agricultural research rose more than 20% between 2001 and 2008. However, most of this growth occurred in only a handful of countries (Beintema and Stads, 2011). In developing countries, funding is often dependent on foreign aid and granted for time-limited projects; this may hamper the development of national R&D institutions and capacity building. However, research in some developed and emerging economies will have spill-over effects to other developing countries. An important challenge is to make research results better adapted to local conditions and to foster the adoption of technologies able to improve productivity growth sustainably in diverse conditions (FAO, OECD et al. [for G20], 2012). Recognising the importance of investment in agricultural research, Development Assistance Committee (DAC) donors called for increased support as part of the L'Aquila Food Security Initiative. Box 2.3 provides information on Official Development Assistance provided for agricultural research for development.

Increasing agricultural land use sustainably

There is less scope for increasing land use than there is for increasing yields. FAO projections to 2050 foresee just 10% of future crop output growth (21% in developing countries) coming from area expansion. This reflects, in part, tightening constraints on global land and water availability; as well as greater optimism about the strong potential for yield growth in some of the poorest regions of the world.

FAO estimates that total arable land will increase, but only by 0.1% per year (less than 4% over 35 years), implying a steady decline in the amount of arable land per person. The Agrimonde Foresight study (Paillard et al., 2011) estimates higher increases, with the amount of crop land expansion by 2050 between 19% and 39% depending on the scenario. Higher yields are expected as a result of technological progress, and investments in agricultural research and irrigation systems.

The analysis of Global Agro-Ecological Zones (GAEZ) (Fischer et al., 2009) suggests that there is little or no room for expansion of arable land in South Asia, the Near East and North Africa. Where land is available, in sub-Saharan Africa, Latin America and Central Asia, more than 70% suffers from soil and terrain constraints (Alexandratos and Bruinsma, 2012). That land is also subject to competition from other uses (urban growth, industrial development, environmental reserves and recreational uses). The competition for competing land use will be resolved according to economic incentives, but those incentives may need to be regulated to ensure sustainable resource use and to address concerns about the social implications of land use changes (e.g. "land grabs").

Land quality is as important as total area. Considerable areas of productive land have been lost through degradation of soil, abandonment or different types of pollution, and restoring this land for cultivation or grazing is a way of increasing food production. The UK Foresight study suggests that land degradation costs an estimated USD 40 billion annually worldwide.

Policies are also important. Improving land tenure systems can have an important effect on famers incentives to look after their land (OECD, 2011a), and is central in ensuring that any change in farm structures occurs fairly. Foresight (2011) and Hertel (2010) stress that public investments in global databases about land use patterns and land quality would help in the design of a rational land use policy. The FAO's *Voluntary Guidelines on responsible governance of tenure of land, fisheries and forests in the context of national food security* (FAO, 2012c), endorsed by the Committee on World Food Security (CFS) in May 2012, outline the basic principles which should govern land tenure reforms designed to ensure sustainable and inclusive land use.

Making more efficient use of scarce water resources

Water is an essential input for agricultural production. At the global level, agriculture accounts for about 70% of total water withdrawal. In some countries, over 90% of withdrawals are for agricultural purposes. Cities and industries are competing intensely with agriculture for use of water, and an increasing number of countries, or regions within countries, are reaching alarming levels of water stress and pollution. Global freshwater resources will be further strained in the future in many regions, with over 40% of the world's population projected to be living in river basins experiencing severe water stress by 2050 (OECD, 2012a).

While the majority of cropland cover is rainfed, irrigated areas are considerably more productive and cover some 16% of the arable land in use, accounting for 44% of all crop production and 42% of cereal production in the world. The shares for developing countries are somewhat higher with 21% of arable land irrigated, accounting for 49% of all crop production and 60% of cereal production (Alexandratos and Bruinsma, 2012). Yet a large proportion of the world's food production is based on unsustainable exploitation of groundwater that at the same time is threatened by increasing pollution by agro-chemicals (OECD, 2010). Climate change will also affect the area and productivity of both irrigated and rainfed agriculture across the globe. Thus, measures to deal with climate variability and improve land and water management practices will be necessary to create resilience to climate change and to enhance water security.

The quality of surface and groundwater outside the OECD area is expected to deteriorate in the coming decades (FAO, OECD et al. [for G20], 2012). Water pollution also stems from inappropriate agricultural practices including poor waste management, such as excess nutrient flows due to overuse of inorganic fertilisers and livestock manure. The increase of agricultural production to meet increased demand for food will further exert pressure on water systems.

People who have better access to water tend to have lower levels of undernourishment. In areas that depend on local agriculture, lack of water can be a major cause of famine and undernourishment. Yet by 2025, it is estimated that 1.8 billion people will be living in countries or regions with absolute water scarcity, and two-thirds of the world's population could be living under water stressed conditions (FAO, 2012d). In vulnerable areas, investment in water management techniques should be considered when promoting agricultural productivity growth (OECD and FAO, 2012).

The priority is to use water as efficiently and sustainably as possible. Ways of improving water management practices include drip-feed irrigation, micro sprinklers and the use of no-till agriculture. It will also be important to invest in water infrastructure, in particular by expanding water supply capacity for irrigated agriculture, building water storage capacities, recycling water, improving irrigation infrastructure and taking measures to limit the impacts of drought and flood disasters. Factors that can encourage private investment in irrigation include defining titles to water rights, which encourage infrastructure maintenance and renewal (OECD, 2010).

In order for water to be used efficiently, it is important to create incentives for farmers and other users that reflect the value of water and the costs of pollution so that water users will tend to use less water (by increasing water use efficiency) and diminish pollution. Market incentives range from water charges to formal or informal trading of water user rights. Some OECD and developing countries (such as China) are now moving towards imposing charges that reflect the costs of supply and scarcity of water. The experience in OECD countries shows that the introduction of water charges has helped lower the quantity of water applied per hectare irrigated, but without leading to an overall reduction in agricultural output or incomes (OECD, 2010; FAO, OECD et al. [for G20], 2012). OECD research also shows that removing policies which intensify production, such as subsidies for inorganic fertilisers and pesticides, can reduce water pressure from agricultural activities.

To address water pollution there are also innovative policy tools, such as water quality trading and agreements between water supply utilities and farmers, which can reduce pollution and water treatment costs. Policies to improve water quality need to take into account the changing behaviour of farmers, the agro-food chain and other stakeholders (OECD, 2012b).

Reducing supply chain losses

There are numerous sources of loss and waste in the food system. On the producer side, those losses can occur during production, post-harvest (in storage or distribution) or while processing. The issue of consumer waste is discussed separately in Section 2.4 on the demand side determinants of food availability, although quantitative studies often combine assessments of producer and consumer losses.

There are considerable food losses in developing countries due to inadequate infrastructure, poor storage facilities, weak technical capacity and under-developed markets. A study undertaken for FAO suggests that these losses (without taking into account waste by consumers) range from 26% to 37% of all production or 114 to 159 kg per person year per capita in South Asian and Sub-Saharan African countries (Gustavsson et al., 2011). That figure compares with a figure of 20% or 185 kg per year capita in Europe and North America. Kummu et al. (2012)[5] estimate that globally about 25% of food produced, corresponding to 614 kcal per person per day is lost. Of that total, just over half is lost on the production side – in the field, post-harvest or during processing. The remainder is lost at the distribution and consumption stage. In terms of natural resources used for food production those losses account for 23% of land, 24% of freshwater resources and 23% of fertiliser. According to this estimate, a 50% reduction in global food losses would produce enough food to feed 1 billion people (Kummu et al., 2012). While there are few studies, and – as the studies' authors note – the findings need to be interpreted with caution, the losses are clearly important.

Waste on the production side can be reduced by improvements in harvest techniques, farmer education, storage facilities and cooling chains, and the development of infrastructure (roads, energy sources and markets). The UK Foresight report suggest that public and donor financing should be directed to locally relevant infrastructure improvements (Foresight, 2011). Better links between smallholders and regional and international food chains (for example by using mobile phones to access information) can improve the consistency and quality of food supply, providing in turn better returns on investment and allowing for reductions in seasonal oversupply and wastage.

Renewable energy and biofuel policies

The use of agricultural crops for ethanol and biodiesel production is having a significant effect on world food markets.[6] The OECD and FAO *Outlook* anticipates that global ethanol and biodiesel production will continue to expand over the coming decade, supported by high crude oil prices and a continuation of policies promoting biofuel use (OECD and FAO, 2012), although the rate of increase will slow. In the longer term, Hertel et al. (Hertel et al., 2012) suggest that if oil prices continue to grow strongly, then biofuel production will continue to expand, even without subsidies or GHG targets. However, there are huge uncertainties about the scale of impact on overall land use, largely because technological developments in biofuels and the availability of fossil fuels are difficult to predict.

At present, the United States, Brazil and the European Union dominate the ethanol and biodiesel markets, while Argentina is also significant in the biodiesel market. Production and use of biofuels in United States and the European Union are driven predominantly by the policies in place. While policies have had an impact in Brazil, the growing use of ethanol is linked to the development of a flex-fuel vehicle industry and, more recently, to policy induced import demand from the United States.

By 2021 the OECD-FAO *Outlook* (OECD and FAO, 2012) projects that 14% of global coarse grains production and 34% of global sugarcane production will be used for ethanol production. About 16% of global vegetable oil production will be devoted to biodiesel production. US ethanol accounted for half the global increase in cereals consumption between

2005/06 and 2007/08 (Westhoff, 2010). Between 2008-11 and 2012-21, the average share of biofuel use in total demand is projected to increase modestly, by 2.6% for coarse grains, 0.8% for wheat and 3.6% for vegetable oils (Figure 2.6). Scenario analysis in the OECD and FAO *Outlook* suggests that narrowing the productivity gap between developed and developing countries could lead to a significant increase in the share of crops that goes into biofuel production (OECD and FAO, 2012).

Figure 2.6. Changes in share of demand increases of several crops 2008-11 and 2012-21

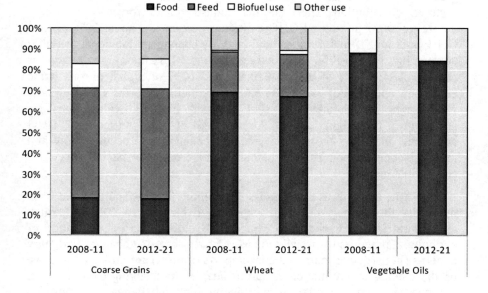

Source: Own elaboration based on OECD and FAO (2012).

Expansion of biofuels would push up prices for many food staples, but there is huge uncertainty over the magnitude of impacts owing to uncertainties over energy prices and policies (Matthews, 2012a). FAO estimates that maize prices could be between 25% and 71% higher by 2050, depending on the scenario (Alexandratos and Bruinsma, 2012). On the other hand, Fischer finds higher cereal price changes of between 20% to 40% by 2020, but lower impacts in 2050 due to the rise of second generation biofuels (Fischer et al., 2009). The price impacts are sensitive to the share that first-generation biofuels are mandated to contribute to total transport fuel consumption.

This analysis supports the IO recommendation to the G20 to remove all policies that subsidise or mandate the production and consumption of biofuels made from raw materials that compete with food, and to open international markets so that renewable fuels can be produced where it is viable to do so. It also underlines the importance of encouraging research into fuels which use feedstocks that do not compete with food (FAO, OECD et al. [for G20], 2012).

Climate change

Agriculture is a major net contributor of GHGs, with nitrous oxide and methane emissions accounting for around 14% of total anthropogenic greenhouse gas emissions (IPCC, 2007) – making it the fourth largest sectoral contribution after energy, industry and forestry (including deforestation). Agricultural GHG emissions account for about 30% of total GHG emissions if fuel, fertiliser and land use change are included, the latter accounting for 6–17% of total emissions. Livestock production is responsible for 37% of global methane and 65% of global nitrous oxide emissions, and 18% of total GHG emissions, including effects through land use change and deforestation (not included in IPCC calculations for agriculture) (Foresight, 2011). About 75% of total agricultural GHG emissions, including those from land use change, now occur in low and middle income economies, and their share is increasing, especially in Africa and Latin America.

Methane and nitrous oxide emissions coming from agriculture and from the wider food supply chain are expected to increase to 2050. Although agricultural land is expected to expand only slowly, the intensification of agricultural practices (especially the use of fertilisers) and changes in dietary patterns (in particular increased consumption of meat) are projected to drive up these emissions. While crops can be adapted to changing environments, the need to reduce emissions will increasingly challenge conventional, resource-intensive agricultural systems (Royal Society, 2009 cited by Foresight, 2011).

In response, a wide range of GHG mitigation measures (for reducing or promoting active carbon sequestration) are likely to be adopted from now until 2050. Market mechanisms, such as carbon taxes, emissions trading and product certification (to incentive changes in consumer behaviour) have the potential to lower emissions, as do selective regulations. However, these measures need to be balanced against the wider challenges of ensuring food availability. Management of land and aquatic systems currently provide the only practical means to enhance the capture and storage of carbon. If water becomes scarcer with climate change, improving water quality by reducing farm emissions will be critical (OECD, 2011b).

Ways of reducing carbon emissions and stimulating carbon sequestration include restoring degraded lands, reforesting; optimising nutrient use by more precise dosage of inorganic fertilisers; improving productivity (output per unit of GHG); reutilising agricultural waste and finally reducing the carbon intensity of fuel and raw material inputs through improvements in energy efficiency and the use of alternative sources (Foresight, 2011). Reducing producer and consumer food losses also implies that less food needs to be produced and therefore less GHG emitting activities need take place.

OECD work on climate change has stressed that those responsible for climate change should bear the costs of mitigation. Governments can put that principle into practice by supporting efficient adaptation programmes that target local sources of climate change (OECD, 2011b). Effective adaptation should significantly reduce the damage resulting from climate change. For example, investment in research, irrigation, rural roads could offset the crop productivity losses driven by climate change (OECD, 2011b). Furthermore, the negative effects of climate change on food security can be counteracted by economic growth, higher agricultural productivity and open international trade in agricultural products to offset regional shortages (Nelson et al., 2009). However, production will ultimately need to migrate from areas where it becomes inherently unsustainable (for example due to chronic or recurring drought and desertification).

The primary role of governments in climate adaptation is to provide public policies that help the private sector adapt. One key area is in providing more accurate assessments of climate change, allowing farmers to make anticipatory changes. Another central role is in research, for example in supporting the development of new seed varieties. Water policies, as well as land use and land management policies, can also be important in providing farmers with incentives to adapt. Government subsidies to weather insurance have been proposed as a possible risk management tool, but induce moral hazard by reducing farmers' incentives to move away from high risk locations.

2.4. Reducing demands that are detrimental to food security

Modifying food preferences

Rising incomes lead to increased calorie consumption, while creating demands for more protein and greater diversity in consumption. Up to a certain point that leads to healthier diets; but beyond that, people tend to consume too many calories and more meat, sugar and vegetable oils than are required for a healthy diet (Figure 2.7).

Figure 2.7. Income growth and dietary changes

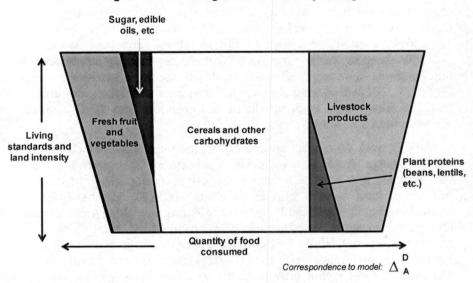

Source: Southgate et al. (2011).

In global terms, average calorie consumption is around 2 800 kcal per person today. This contrasts with a minimum daily energy requirement (MDER) that, taking into account age structure and activity levels, averages about 1 850 kcal per day across countries (the figure is about 2 100 per adult). However, inequality of access implies that about one in seven people are undernourished, even with this surplus of calorie availability. For most countries, existing income inequalities imply that about 2 800 kcal per day is in fact the average consumption level needed to ensure that no more than 1-2% of the population falls below the minimum daily energy requirement.

There are at least as many people over-nourished in the world as are under-nourished, and over-nourishment is an increasing health problem in developing countries. Physically redistributing food from the former to the latter is not a realistic proposition, but modifying dietary patterns so that people have healthier diets would reduce overall demand and prices, and associated pressure on land and water resources. The option of using food taxes to constrain excessive demand has to take account the regressive nature of such measures and the fact that most foodstuffs – unlike say tobacco – are good for health within limits. The most straightforward first step to reducing excessive consumption is through the provision of information and education which can change food consumption cultures.

Figure 2.8. Developing countries: Population with given kcal per person per day

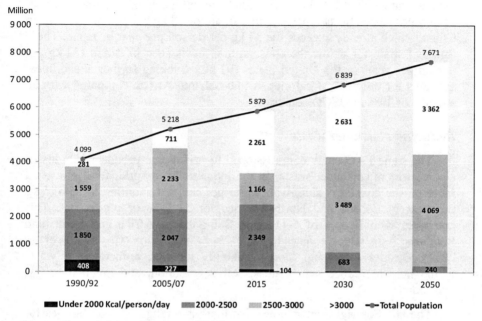

Source: Alexandratos and Bruinsma (2012).

High levels of meat consumption are a particular contributor to excessive calorie consumption (and overall to excessive fat consumption too), with many developed countries consuming more meat than is recommended by nutritionists. Moreover, meat consumption exerts a strong demand on land and water resources. It takes 2 tonnes of grain to produce a tonne of poultry, four tonnes of grain to produce a tonne of pork, and between seven and ten tonnes to produce a tonne of beef. Lower meat consumption would enable more of that grain production to be allocated directly to food use. Lower consumption of sugar, which is weak in nutritional value, would similarly allow resources to shift into other crops.

FAO expects that whereas currently 53% of all the calories consumed in developing countries are provided by cereals and 20% by meat, dairy and vegetable oils, by 2050 the contribution of cereals will have dropped to 47% and that of meat, dairy and fats will have risen to 29% (Figure 2.8).

A major factor behind recent changes in demand has been rapid growth in the consumption of livestock products in countries like China and Brazil. As incomes rise, changes in meat consumption will have potentially important implications for food availability and land use (Paillard et al., 2011). However, there are many uncertainties over how meat demand will respond to income growth, with that response dependent on a range of factors, some cultural. India's meat consumption is low, at less than 4 kg per person per year. This compares with a figure of 48 kg per person per year in China, which already exceeds the 34 kg per person per year in Japan. These figures remain far below the consumption levels seen in Brazil (84 kg per person per year) or the United States (91 kg). Looking further ahead, meat accounts for only 6% of calories in Sub-Saharan Africa, compared with an average of 30% in OECD countries.

Reducing consumer waste

The same FAO study commissioned to investigate producer-side losses also considers consumer waste. In the industrialised world, food is wasted more on the consumer side, with waste per-capita amounting to 95-115 kg per year in Europe and North-America, or 11-13% of production. This compares with figures of 6-11 kg in Sub-Saharan Africa and South and Southeast Asia, which equates to just 1-2% of production (Gustavsson et al., 2011). Besides increasing food availability directly, reductions in waste would help to reduce water stress, soil degradation, and greenhouse gas emissions.

The UK Foresight report notes that there is a range of opportunities for reducing consumer and food service sector waste such as public campaigns, advertising, taxes, regulation, purchasing guidelines and improved labelling (Foresight, 2011). One suggestion is that commercial and charity organisations could arrange for the collection and sale or use of discarded "sub-standard" products that are still safe and of good taste and nutritional value (Gustavsson et al., 2011). A significant share of production is wasted because it does not meet standards for shape or appearance. Raising awareness among food industry, retailers and consumers is needed to reduce these and other forms of waste (OECD and FAO, 2012).

Notes

1. There would be an aggregate global impact resulting from the accumulated changes to food supply and food demand, with direct effects on domestic prices for countries that are not integrated with world markets.

2. www.agmip.org.

3. There is some ambiguity in comparing trends because output measures for different crops and livestock products can be aggregated using different units – mass-based, calorie-based and price-based. Tilman et al. use a calorie-based measure but confine themselves to crop demand, although growth in livestock consumption is implicitly accounted for by taking into account the use of crops for feed. Calorie consumption oversimplifies the challenge because of a trend towards greater diet complexity. Staple food consumption will increase more slowly than calories, but consumption of meat, sugar, oils, fruits and vegetables will grow more rapidly.

4. Climate change is leading to rising temperatures. The IPCC anticipates that global temperatures will rise by between 1.50 and 4.50 by 2100 (10 to 30 by 2050) (IPCC, 2007). It also involves other changes to nature that affect agricultural production potential, including to radiation, rainfall, and soil and water availability. In addition sea levels are expected to rise, leading to salt water inundation and intrusion along coastlines, while extreme weather events (e.g. droughts, floods, thunderstorms and heat waves) may become more frequent or intense, posing a significant challenge to food security.

5. Their calculation uses the loss and waste percentages of Gustavsson et al. 2011

6. Analysis of indirect land use change has fundamentally altered assessment of the impacts that biofuels have on greenhouse gas (GHG) emissions. Previously, biofuels were considered as an instrument to reduce GHG emissions, but recent research suggests that over the decades to 2050 and perhaps beyond, GHG emissions could rise due to biofuel expansion, mainly because of destroyed pasture and forest areas. However, that analysis also finds cumulative GHG emissions turning negative later in the century as second generation biofuels come on-stream (Hertel et al., 2012).

References

Alexandratos, N. and J. Bruinsma (2012), "World agriculture towards 2030/50: The 2012 revision", *ESA Working Paper* No. 12-03, FAO, Rome.

Alston, J. (2010), "The benefits of agricultural research and development, innovation and productivity growth", *OECD Food, Agriculture and Fisheries Working Papers*, No. 31, OECD Publishing, Paris.

Beintema, N. and G. Stads (2011), "African agricultural R&D in the new millennium: Progress for some, challenges for many", *Food Policy Report* 24, International Food Policy Research Institute, Washington, DC.

Bruinsma, J. (2011), "The resource outlook to 2050: By how much do land, water use and crop yields need to increase by 2050?", Chapter 6 in *Looking Ahead in World Food and Agriculture: Perspectives to 2050*, Conforti, P. (ed.) 2011, FAO, Rome.

Evenson, R. and K. Fuglie (2010), "Technological capital: The price of admission to the growth club", *Journal of Productivity Analysis* 33 (3), pp. 173-190.

Fan, S. and X. Zhang (2008), "Public expenditure, growth and poverty reduction in rural Uganda", *African Development Review* 20(3), pp. 466-496.

Fan, S., B. Yu and A. Saurkar (2008), "Public spending in developing countries: Trends, determination and impact", in *Public Expenditures, Growth and Poverty*, Fan, S. (ed.), John Hopkins University Press, Baltimore.

FAO (2012a), *The State of Food Insecurity in the World*, FAO, Rome.

FAO (2012c), *State of Food and Agriculture. Investment in Agricultural for Food Security*, FAO, Rome.

FAO (2012d), *Voluntary Guidelines on the Responsible Governance of Tenure of Land, Fisheries and Forests in the Context of National Food Security*, FAO, Rome.

FAO (2012e), *FAO Water, Natural Resources and Environment Department*, www.fao.org/nr/water/issues/scarcity.html, FAO, Rome.

FAO (2011), *Climate Change, Water and Food Security*, FAO, Rome.

FAO, OECD et al. [for G20] (2012), "Sustainable agricultural productivity growth and bridging the gap for small-family farms", Interagency Report to the Mexican G20 Presidency with Contributions by: Bioversity, CGIAR Consortium, FAO, IFAD, IFPRI, IICA, OECD, UNCTAD, UN, WFP, World Bank, WTO, G20 Mexican Presidency.

Fischer, G. (2009), "World food and agriculture to 2030/50: How do climate change and bioenergy alter the long-term outlook for food, agriculture and resource availability?", Presented at the Expert Meeting on How to Feed the World in 2050, 24-26 June, FAO, Rome.

Fischer, G., H.v. Velthuizen and F. Nachtergaele (2009), *Global Agro-ecological Assessment, 2009*, revised edition, International Institute for Applied Systems Analysis, Laxenburg.

Foley, J., A.N. Ramankutty, K.A. Brauman, E.S. Cassidy, J.S. Gerber, M. Johnston (2011), "Solutions for a cultivated planet", *Nature*, 478, pp. 337-342.

Foresight (2011), *The Future of Food and Farming: Challenges and Choices for Global Sustainability*, The Government Office for Science, London.

Fuglie, K. (2012). "Productivity growth and technology capital in the global agricultural economy", in Fuglie, K.O., S.L. Wang, and V.E. Ball (eds.), *Productivity Growth in Agriculture: An International Perspective*, CAB International, Oxfordshire.

Gustavsson, J., C. Cederberg, U. Sonesson, R. van Otterdijk and A. Meybeck. (2011), "Global Food Losses and Food Waste: Extent, Causes and Prevention", Study conducted for the International Congress *Save Food!*, Düsseldorf, FAO, Rome.

Hertel, T.W. (2010), "The global supply and demand for agricultural land in 2050: A perfect storm in the making?", AAEA Presidential Address, *GTAP Working Paper* No. 63, West Lafayette.

Hertel, T., J. Steinbuks and U. Baldos (2012), "Competition for land in the global bioeconomy", *GTAP Working Paper* No. 68, Presented at the meeting of the International Association of Agricultural Economists, Foz do Iguacu, Brazil, August 18-24, 2012.

International Sustainability Unit (2011), *What Price Resilience? Towards Sustainable and Secure Food Systems*, ISU, London.

Kummu, M., H. de Moel, M. Porkka, S. Siebert, O. Varis and P.J. Ward (2012), "Lost food, wasted resources: Global food supply chain losses and their impacts on freshwater, cropland, and fertiliser use", *Science of the Total Environment*, 438, pp. 477–489.

Ludena, C., T.W. Hertel, P.V. Preckel, K. Foster and A. Nin (2007), "Productivity growth and convergence in crop, ruminant and nonruminant production: Measurement and forecasts", *Agricultural Economics*, 37(1), pp. 1-17.

Malthus T.R. (1798), *An essay on the principle of population*, J. Johnson, London.

Matthews, A. (2012a), "Global food security and the challenges for agricultural research", Presentation to the Conference 'Innovation and competitiveness of the agrarian sector of the EU', Prague, 17 September.

Mueller, N. D., J. S. Gerber, M. Johnston, D. K. Ray, N. Ramankutty and J.A. Foley (2012), "Closing yield gaps through nutrient and water management", *Nature*, 490, pp. 254-257.

Nelson, G., M. Rosegrant, A. Palazzo, I. Gray, C. Ingersoll, R. Robertson, S. Tokgoz, T. Zhu, T.Sulser, C. Ringler, S. Msangi, and L. You (2010), "Food security, farming, and climate change to 2050: Scenarios, results, policy options", *IFPRI Issue Brief* No. 66, IFPRI, Washington, DC.

OECD (2012a), *Farmer Behaviour, Agricultural Management and Climate Change*, OECD Publishing Paris.

OECD (2012b), *OECD Environmental Outlook to 2050:The Consequences of Inaction*, OECD Publishing, Paris.

OECD (2011a), *Agriculture and Economic Adaptation to Climate Change*, OECD Publishing, Paris.

OECD (2011b), *Fostering Productivity and Competitiveness in Agriculture*, OECD Publishing, Paris.

OECD (2010), *Sustainable Management of Water Resources in Agriculture*, OECD Publishing, Paris.

OECD (2009), *The Bioeconomy to 2030: Designing a Policy Agenda, Main Findings and Policy* Conclusions, OECD Publishing, Paris.

OECD (2003), *Farm household income: Issues and policy responses*, OECD

OECD and Eurostat (2005), *Oslo Manual Guidelines for Collecting and Interpreting Innovation Data*, OECD Publishing, Paris.

OECD and FAO (2012), *OECD/FAO Agricultural Outlook 2012-2021*, OECD Publishing, Paris and FAO, Rome.

Paillard, S., S. Treyer and B. Dorin (2011), *Agrimonde Scenarios and Challenges for Feeding the World in 2050*, Editions Quae, Versailles.

Rosegrant, M. et al. (2012), "Water and food in the bioeconomy-challenges and opportunities for development", Plenary Paper prepared for presentation at the International Association of Agricultural Economists (IAAE) Triennial Conference, Foz do Iguaçu, Brazil, 18-24 August, 2012.

Sen, A. (1980), "Famines", *World Development* 8(9), pp. 613-621.

Southgate, D., H. Graham, L. Tweeten (2011), *The global food economy*, Blackwell Publishing, Oxford, UK.

Tilman, D. C. Balzer, J. Hill and B. L. Befort (2011), "Global food demand and the sustainable intensification of agriculture", *Proceedings of the National Academy of Sciences of the United States of America*, 108(50), pp. 20260-20264.

Tubiello, F., and G. Fischer (2007), "Reducing climate change impacts on agriculture: Global and regional effects of mitigation, 2000-2080". *Technological Forecasting and Social Change*, 74, 7, pp. 1030-1056.

Tweeten, L. and S. Thompson (2008), "Long-term global agricultural output supply demand balance and real farm and food prices", *Working Paper* 0044-08, Department of Agriculture Economics, Ohio State University, Columbus

Westhoff, P. (2010), *The Economics of Food*, FT Press, Upper Saddle River, New Jersey.

Wiggins, S. and R. Slater (2010), "Food security and nutrition: current and likely future issues", *Science Review* 27, Foresight Project on Global Food and Farming Futures, Government Office for Science, London.

World Bank (2012a), *Global Monitoring Report 2012: Food Prices, Nutrition, and the Millennium Development Goals*, International Bank for Reconstruction and Development/The World Bank, Washington, DC.

Chapter 3

The role of food and agricultural trade in ensuring domestic food availability

This chapter examines the role of agricultural trade in ensuring that food is available domestically. It considers the balancing role of international and regional trade, the benefits and costs associated with open markets, as well as the ways in which governments can manage shocks emanating from both domestic and international markets.

Food security is dependent upon food being available on national markets, whether that food is produced domestically or in other countries. This section considers the role of trade in ensuring stable food availability. Specifically, the first section looks at the balancing role of trade in reallocating food from surplus to deficit areas, while the second section explores the specific role of regional trade. The issue of developing countries' food import bills and whether dependence on food imports raises question regarding affordability is the considered, while the final section examines the links between trade and the stability of domestic food supplies.

3.1. The balancing role of trade

Trade plays a vital role in balancing the deficits of net food importers with the surpluses of net food exporters. In the absence of trade, food prices would be higher in current net food importers in order to bring national supply and demand into equilibrium, worsening the food security status of consumers in those countries. In the absence of trade, food prices would be lower in net exporting countries because of the inability to export surplus production.

The relative importance of agricultural trade to domestic food availability remains small, even if at the margin it is hugely important for those countries that rely on it. In most countries, domestic production is the main source for domestic consumption and trade plays a relatively minor role. For the world as a whole the share of trade relative to production differs across commodities (Liapis, 2012). Measuring the trade share in terms of the quantity of exports (thus avoiding problems due to varying prices over time) and comparing the ratio among the selected commodities, rice had the lowest share of production exported while whole milk powder had the highest. Liapis (2012) also shows that in most cases the export share of production has not changed dramatically over recent decades. Rice, sugar, whole milk powder and soybean oil have experienced rising export shares, shares for maize and butter have declined, and there is no discernible trend in the shares for wheat, soybeans and beef.

Another perspective on the role of agricultural trade in contributing to food availability is provided by focusing on countries rather than commodities (Figure 3.1). The conventional view has been that developing countries are net agricultural exporters and developed countries are net agricultural importers. This was indeed the case, but during the 1960s and 1970s, the developed country share of world agricultural exports rose and its share of world agricultural imports fell, while for developing countries the reverse was true. Since about 1980 both groups of countries have been in an approximate balance, although developing countries' share of world trade has risen since the 1990s. By 2010 developed country export (and import) shares had fallen to 60% of the world total, while developing country export (and import) shares had risen to 40% of the total.

Figure 3.1. Developed and developing country shares of world agricultural trade

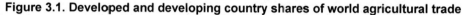

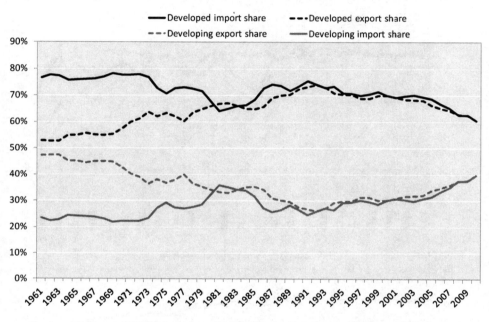

Source: Matthews (2012b) based on FAOSTAT. Developing countries include transition economies.

The aggregate trends in Figure 3.1 do not tell the full story of the structural changes in world agricultural trade. First, the developing country agricultural trade balance is heavily influenced by the phenomenal export performance of Brazil (Figure 3.2). When Brazil is excluded from the developing country aggregate, the sharp deterioration in the net agricultural trade balance of the remaining developing countries becomes clear. The right-hand panel of Figure 3.2 shows that this deterioration in the agricultural trade balance occurred in all three important developing country groups (least developed countries [LDCs], low-income food deficit countries [LIFDCs] and net food-importing developing countries [NFIDCs]).[1]

Figure 3.2. Net agricultural trade of selected developing country groups, 1961-2010

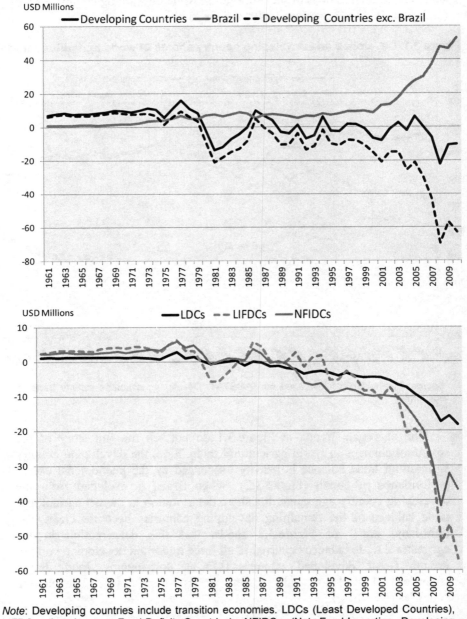

Note: Developing countries include transition economies. LDCs (Least Developed Countries), LIFDCs (Low-Income Food-Deficit Countries), NFIDCs (Net Food-Importing Developing Countries).
Source: Matthews (2012b) based on FAOSTAT.

Second, a number of developing countries are net agricultural exporters but also net food importers (Valdés and Foster, 2012).[2] There has been a steady increase in the number of developing countries which have turned from being net food exporters to net food importers (Table 3.1).[3] In the early 1980s, less than 60% of developing countries were net importers of food; by the late 2000s this proportion had increased to nearly 80%. Whether or not greater dependence on food imports implies a greater risk of food insecurity depends on whether the change reflects a shift of resources from food production into more remunerative activities (and the fact that the opportunity cost of importing food is lower than the opportunity cost of producing it domestically) or is a result of fundamental development failure. Which is the case should be reflected in income levels and the availability of foreign exchange. The highest proportion of net food importers are found in the low-income and high-income groups, respectively. For all countries that depend on food imports, the reliability of the world market as a source of supplies is essential for food security.

Table 3.1. Food trade status of developing countries, by income class

Row labels	1980-85		2005-10	
	Number of countries	Per cent of income group	Number of countries	Per cent of income group
Low-income	36		36	
Net food importer	24	66.7%	33	91.7%
Net food exporter	12	33.3%	3	8.3%
Lower-middle income	49		49	
Net food importer	26	53.1%	34	69.4%
Net food exporter	23	46.9%	15	30.6%
Upper-middle income	50		50	
Net food importer	22	44.0%	36	72.0%
Net food exporter	28	56.0%	14	28.0%
Developing high-income	15		15	
Net food importer	13	86.7%	15	100.0%
Net food exporter	2	13.3%	-	-
All developing countries	150		150	
Net food importer	85	56.7%	118	78.7%
Net food exporter	65	43.3%	32	21.3%

Source: Matthews (2012b) based on FAOSTAT series of exports and imports of food excluding fish. Income classification based on the World Bank (2012) income classification.

The net trade status of countries may also be an outcome of distortionary or discriminatory policies, with countries that protect their agricultural sectors producing more and importing less (possibly exporting) than would be the case if resources were allocated in line with their comparative advantage. This has tended to be the case in the majority of OECD countries, although the degree of protection and other forms of trade-distorting support has declined significantly over the past 25 years (Box 3.1). However, comparative advantages in food production are not irreversibly fixed. While heavily influenced by initial endowments of land, water, climate, topography, soils and the prevalence of pests and diseases, whether a country is a net food exporter or importer also depends on its past investments in agriculture-related physical and human capital, in institution-building and in technology improvement. Some importing countries have neglected investments that improve agricultural productivity, which has increased their import dependence. While the current pattern of agricultural trade may reflect some of these failings, this does not negate the insight that the balancing role of trade makes a fundamental contribution to global food security and to food security in developing countries in particular.

Whether trade is likely to play a more or less important role in balancing food supply and demand in developing countries in future will depend on food demand trends and domestic supply capacities, which evolve at differing rates across countries. There are many reasons why current food importers might expect to experience a steadily deteriorating comparative advantage in food production. Net food importing countries have, in general, more rapidly growing populations and more rapidly growing food demand per capita than net exporters. Net food importers also have, on average, poorer resource endowments in terms of land and water availability, with yield performances that will potentially be more adversely affected by climate change. On the other hand, greater investments in increasing agricultural productivity could significantly affect production and productivity growth. For example, yield gaps are high in many net food importing countries in Africa, and closing these yield gaps would narrow the difference between consumption and production.

Model simulations which capture the combined impact of these supply and demand drivers produce a wide range of estimates of the likely net trade positions of countries in 2050. On balance, they suggest that trade is likely to become increasingly important as a supplement to domestic production in ensuring adequate food availability (and as a source of export earnings and income as will be seen in the following section). For example, the latest FAO projections to 2050 envisage the net cereals imports of developing countries increasing from 116 million tonnes in 2005-07 to 168 million tonnes in 2030 and 196 million tonnes in 2050 (Alexandratos and Bruinsma,

2012). The Global Harvest Initiative's 2012 GAP report compares the growth in projected demand in each region with the growth in projected supply based on a continuation of current growth rates in agricultural total factor productivity. It concludes that, on this business-as-usual scenario, 74% of the growth in total demand can be met by maintaining the current TFP growth rate, leaving a significant gap to be met by imports. In South and South-east Asia, the proportion is 82% and in the Middle East and North Africa it is 83%. For Sub-Saharan Africa, only 13% of the increase in total demand would be met from domestic production on current trends, while Latin America would have a growing exportable surplus (Global Harvest Initiative, 2012). While these projected outcomes can be influenced by policy interventions, they suggest that the balancing role of trade in contributing to food availability in developing countries will become more, rather than less, important over time.

Box 3.1. The evolving effects of OECD country agricultural policies on developing countries

For decades, the agricultural policies of OECD countries have been considered to be thwarting the development prospects of poorer countries. This is because of the high degree of support provided to farmers, and the potentially damaging spill-overs of that support onto developing countries. In recent years, there have been important changes in the level and composition of support, and in the types of policy spill-over effects that are of greatest concern.

In the early years of the Uruguay Round of trade negotiations, OECD countries provided a high degree of support to their farmers, with government transfers accounting for 37% of gross farm receipts (the %PSE) across the OECD area in 1986-88. Moreover, a large share of that support (over 80%) was linked to output, mostly in the form of higher prices than those prevailing on world markets. This in turn required the use of trade policy instruments, which were seen to have a range of negative impacts on developing countries:

- High tariffs on agricultural products, typically several times above those levied on industrial goods, restricted market access for developing country farmers with export potential.

- Elevated prices led to the accumulation of surpluses, which were subsequently "dumped" on developing country markets with the use of export subsidies (sometimes badged as food aid). This undermined local markets for developing country farmers competing with imports.

- Price supports and subsidies, by stimulating production, suppressed prices on world markets, again lowering returns to developing country farmers.

These impacts implied weaker terms of trade for developing countries that were specialised in agriculture.

The nature of the effects of many OECD agricultural policies has not changed but the magnitude has. Price supports and other distorting policies such as output and input subsidies still lead to restrictions in market access, and depress world prices relative to what they would otherwise be. However, the spill-over impacts have become less important because of declines in the rate of support, and changes in the extent to which that support is provided through trade-distorting instruments. Export subsidies have been used only lightly in recent years. By 1999-2001 the share of transfers in gross farm receipts (%PSE) had declined to 32%, with more than two-thirds of that support linked to output. In the past ten years, those changes have accelerated, such that over the period 2009-11, the %PSE averaged 20%, with 45% of transfers linked to output. Recent improvements have been facilitated by stronger world prices, which enable a given domestic price to be maintained with lower support.

The welfare impacts of OECD country policies on developing countries come via efficiency losses and terms of trade effects (which create both winners and losers). The last major OECD effort to calculate these impacts globally was in 2006 when prices were relatively low. In general, the price depressing effects of OECD country policies – calculated when support was considerably higher than it is now – were found to be relatively small for most products, with a 50% cut in all forms of support causing cereal and meat prices to be 2-3% higher than they would otherwise be, and prices for oilseeds and oilseed meal to decline slightly. Dairy products were a notable exception, with 50% cuts causing prices to increase by 13%. These findings were broadly in line with those of other studies conducted around that time (OECD, 2006).

In terms of the overall welfare impacts (calculated using a version of the GTAP model), the main conclusion was that OECD countries should reform primarily because it was in their own interests to do so – in fact they would reap 90% of the benefits from global agricultural reforms. The OECD study noted that the welfare effects of reform on developing countries were complex and would vary by country. Specifically, competitive suppliers would gain from more open markets and from commodity price increases, while net importers of agricultural commodities would lose in the absence of corresponding increases in the prices of goods they export. Some countries also stood to lose from the erosion of benefits of preferential trading arrangements with OECD countries. On balance, OECD analysis concluded that most developing countries would gain from OECD country liberalisation, although the gains were small relative to the benefits of reforming their own policies. Moreover, a large share of the gains were concentrated among a few emerging economy exporters, in particular Brazil (OECD, 2006).

In 2007-08, world food markets were exposed to a severe shock, with world prices for major food staples showing their biggest increase in real terms since the 1970s. Those price changes exceeded by an order of magnitude the price changes that models such as Aglink suggested would flow from OECD country reforms. There was swift recognition that while strong prices offer long term benefits for farmers, the short to medium term impacts on poor consumers are predominantly negative. The current emphasis on the harm that high prices inflict on developing country consumers, as opposed to the harm that low prices inflict on farmers with net sales, has led to charges of inconsistency being levelled at international organisations in general (e.g. Swinnen, 2010), although OECD was always careful to note that price changes in either direction create winners and losers.

The observation that OECD policies to support domestic prices are harmful, partly because they suppress international prices, does not imply that output-linked policies are now to be recommended because they contain upward pressure on world prices. Distortionary policies are inefficient as well as being inequitable in terms of their domestic effects (OECD, 2001; OECD, 2003), and globally they prevent resources from being allocated in an efficient way – even if concerns about the pattern of winners and losers have shifted compared to the period when prices were low.

Higher prices have, however, caused the issues to change. The price depressing effects of OECD countries' policies are no longer the immediate concern and the use of export subsidies has almost disappeared. The agricultural policies of most immediate concern are those that contribute to higher and more volatile world prices, namely export restrictions by exporters, temporary tariff reductions by importers, and government support for diverting crops to biofuel production. The use of these instruments is not confined to OECD countries. Export restrictions – which are only weakly constrained by WTO rules – were used mostly by emerging economies in 2007 and 2008 (Jones and Kwiecinski, 2010). Biofuel policies in the United States and the European Union affect mostly the grains and oilseeds sector, but Brazil's hugely important ethanol sector uses mainly sugar cane and could in principle thrive without support policies. Government support policies make world market prices of these products (and their substitutes) substantially higher than they would be, while mandates add to price volatility by creating demand that is less responsive to prices.

More generally, developing countries, in particular the BRIICS, are becoming increasingly important to world agricultural trade. Whereas trade between OECD countries accounted for 58% of world agricultural trade in 1999, by 2009 that share had fallen to less than half. Brazil is now the third largest agricultural exporter in the world, after the European Union and the United States, with more than USD 50 billion of agricultural exports per year. China is simultaneously the fourth largest exporter and the fourth largest importer (with a net deficit), exporting labour intensive products and importing land intensive products in line with its comparative advantage.

As developing countries become richer, and more important to international trade, it is essential to look at a wider web of interactions and policy effects. In particular, the developed (OECD) versus developing country distinction is becoming a less and less relevant lens through which to view the links between agricultural policies and spill-over impacts onto developing countries.

Source: OECD (2012a).

3.2. The importance of regional food and agricultural trade

Regional trade has the potential to improve food security, especially in countries where deeper integration with world food markets remains difficult. Well-functioning regional markets can reduce the cost of food, its volatility and the uncertainty of supply. The major benefit of intra-regional trade is to link food surplus areas with food deficit areas, particularly for food staples. Increased regional trade can boost agricultural growth in surplus zones while mitigating shortages in deficit ones. Studies in Sub-Saharan Africa, for example, show that prices for maize and cassava fall significantly when there are open borders (Dorosh et al., 2009).

In many regions, rural food surplus production zones supply major deficit urban consumption centres as their natural markets, but the presence of borders often adds significant costs to moving food within these natural 'food sheds' (World Bank, 2012b). For example, staple foods trade regularly across national borders in Eastern and Southern Africa. Principal maize surplus areas lie in South Africa, Northern Mozambique, Southern Tanzania and Eastern Uganda and to a lesser extent in Northern Zambia and Northern Tanzania. Sourcing supplies from these surplus areas, local traders supply deficit markets in Southern Mozambique, Malawi and Kenya (Dorosh et al., 2009). Nonetheless, cross-border trade in food staples in Africa remains limited, and prices for staples, especially in land-locked countries, can vary substantially between years of domestic good harvest and those of poor harvest (World Bank, 2012a).

Cross-border trade flows can also potentially help to reduce price volatility in staple food markets where countries in a region are affected differently by exogenous shocks such as weather. Different seasons and rainfall patterns and variability in production, which will increase as climate change continues, imply variable market conditions across countries. Where production variability is not highly correlated among most countries in the region, integration through regional trade offers the prospect of cancelling the effects of small country size on production volatility (Koester 1986). Studies have calculated the amount of stocks needed for each country within a defined region so as to stabilise cereal consumption in times of fluctuations in cereal production and import prices (Dorosh et al., 2009; Koester, 1986). These studies have compared those stock levels to the levels required by the same countries when co-operating regionally. Their results show regional stocks to be more efficient than the sum of national stocks without co-operation.

There are significant differences in the importance of intra-regional agricultural trade in different regions, even taking account of differences in region size, country size and the overall value of trade. Among developing country regions, more than half of Asia's agricultural exports go to other Asian countries, but the intra-regional shares drop to 17% in South and Central America and 20% in Africa (WTO, International Trade Statistics for 2011). However, food trade is largely informal in a number of regions, which means that the importance of regional trade may be significantly underestimated. A recent study by the OECD's Sahel and West Africa Club suggests that regional staple trade in West Africa could be two to three times bigger than official estimates (SWAC, 2012).

The constraints to increasing intra-regional trade are sometimes the lack of physical infrastructure, such as roads, but more often a consequence of government interventions. A recent study of agricultural supply chains in Central America shows that between 29% and 48% of the import prices of grains comes from logistics costs (World Bank, 2012b). In Africa, poor infrastructure was seen as the main impediment to intra-African trade, with transport costs accounting for 50–60% of total marketing costs (GTZ, 2010). However, the road infrastructure along the major international trade corridors has improved, and the high costs are often due to difficulties with "soft" infrastructure, such as roadblocks and licensing arrangements which limit competition among transport operators. Export bans, unnecessary permits and licenses, costly document requirements, and conflicting standards raise transaction costs, add to uncertainty and often lead to the exit of private sector traders from participation in regional trade.

Regional trade agreements (RTAs) can help to facilitate intra-regional trade if they address these barriers and help to create a more predictable trade environment. The share of duty free tariff lines in South-south agreements is expected to increase from 28% to approximately 92% when fully implemented, while North-South agreements increase their share of duty free lines from over 68% to only 87% (Fulponi et al, 2011).

Nonetheless, these agreements have often failed to live up to their potential in practice, with governments often ignoring their obligations and resorting to unilateral measures when convenient. For example, cereal markets across South Asia (especially Bangladesh-India rice and Pakistan-Afghanistan wheat) are increasingly connected. As a result, cereal price policies have major spill-over effects across borders. The 2007-08 experience with the surge in international market cereal prices illustrated some of the shortcomings of these trading relationships, as export restrictions by India and Pakistan contributed to higher prices for consumers in Bangladesh and Afghanistan (Dorosh, 2008). Liberalising agricultural trade within a regional agreement needs strong political will and countries have to be prepared to give up some autonomy in designing and implementing their domestic food policies. With governments committed to pursue interventionist policies to stabilise prices, regional trade flows are often sacrificed when they appear to stand in the way of national food security – despite the longer-term costs.

Increasing regional trade flows requires enabling public sector actions across a number of fronts: investment in physical infrastructure where needed; regulatory reform; improvement of customs administration; harmonisation of standards; and greater transparency regarding trade policy. In many countries, the absence of a stable and predictable policy environment at the national and regional level has created mistrust between government and the private sector (World Bank, 2012b). Effective communication and interaction at the national level between national ministries, the business sector and civil society is required to build understanding and support for the role of intra-regional trade.

3.3. Food import bills of developing countries

More recent concerns with the food security implications of agricultural trade have focused on the growing net food import status of developing countries and the sustainability of sharp rises in their food import bills. According to FAO (2012a), global food import bills have exceeded one trillion dollars each year in the three years 2010-12. As noted earlier, developed countries account for the largest share of food imports (USD 777 billion in 2012 compared to USD 466 billion for developing countries). However, food import bills are rising more rapidly in developing countries. Excluding fish, food import bills in developed countries increased by 240% between the early 1990s and the late 2000s. The equivalent percentage increases for developing countries as a whole were 370%, for least developed countries 385% and for the LIFDC group, 466%.

These figures are sometimes interpreted as a sign of growing dependence on the world market for basic food supplies.[4] However, this conclusion is not necessarily warranted. Part of the increase in food import bills is due to higher prices rather than greater import volumes. Konandreas (2012) estimates that the net cereal bill of NFIDCs almost quadrupled between the mid-1990s and 2010, while their import volumes increased by around 70% – although there was considerable variation in country experiences. Other data in Konandreas (2012) show that self-sufficiency in cereals has remained remarkably stable in both LDCs and NFIDCs over the past thirty years, at 90% and 70% respectively. Nonetheless, constant self-sufficiency ratios are consistent with a growing volume of commercial cereal imports. Paying for food imports can strain the resources of countries where economic growth lags and foreign exchange earnings are limited. Yet despite the sharp rise in both food prices and food import bills in recent years, food-importing developing countries on aggregate experienced remarkable stability in financing food imports during this period, although with very different experiences across countries.

Food import bills in developing countries have sometimes been compared to GDP as an indicator of affordability. This indicator may have some significance as an indicator of trade openness, but what counts is the ability of countries to pay for these imports. Increased food import dependence can be a natural trend during the transition of an agrarian economy to one based more on manufacturing and services, and can be managed if the export income generated by the non-food sectors can pay for the increased food imports. Such trends can result from the successful transition to more productive and diversified structures, and can be accompanied by increased agricultural productivity. Mellor and Johnston (1984), based on Mellor (1966), showed how countries with high rates of agricultural growth can also have large and increasing food imports. They noted that the 16 developing countries with the fastest growth in staple food production during the period 1961-76 also more than doubled their net imports of food staples during that period. In many LDCs, however, growing dependence on food imports seems to suggest more a failure of agricultural development.[5]

Two indicators throw light on the ability of developing countries to finance food imports. The first is the share of food import expenditure in total merchandise imports. A rising share might suggest increasing difficulty in acquiring the desired level of imports. Figure 3.3 shows food import shares expressed relative to total merchandise imports over the period 1961-2010 for a number of developing country groups. For the world as a whole, the importance of food imports in merchandise imports is falling, from around 15% in 1961 to around 5% today. The shares for LIFDCs and for SIDSs follow broadly the same trends. For example, the food import share for LIFDCs was 20% in 1961 but had fallen to 5% in 2010. Trends for NFIDCs and LDCs are less favourable, mainly because shares remained stable between 1961 and the mid-1980s and only began to decline at that point. The experience of the different groupings during the 2008-10 food price spike has been mixed. Shares remained roughly stable for LDCs but increased slightly for the other three groupings. This evidence suggests that, for food-importing developing countries in general, meeting the cost of food import bills has become less onerous over time. Konandreas (2012) conducts a more disaggregated analysis for the period 1990-2009, looking at individual countries in the LDCs and NFIDC groups and at their average, maximum and minimum shares of food import expenditure in both total merchandise imports and exports. His analysis emphasises the volatility of these shares for individual countries from year to year rather than the trend over time. He finds the share of food and animal products in the aggregate merchandise imports of LDCs averaged 17% (simple average) and varied modestly around that level (1990-2009 period), but that of individual countries averaged as much as 42% and for some years reached a maximum of over 80%. For NFIDCs, the situation is not as dramatic, with an average of food imports to total merchandise imports of about 12% and as much as 18% for some countries. Also, for this latter group, the maximum share experienced by any NFIDC country was less than 30%.[6]

Figure 3.3. Food imports as share of total merchandise imports, 1961-2010

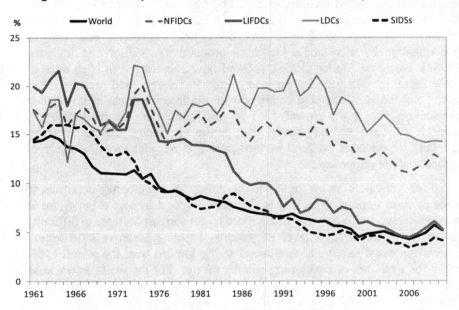

Note: NFIDCs (Net Food-Importing Developing Countries), LIFDCs (Low-Income Food-Deficit Countries), LDCs (Least Developed Countries), SIDCs (Small Island Developing States).
Source: FAOSTAT.

A second indicator of affordability is the coverage ratio, defined as the share of food import expenditure in a country's foreign exchange earnings. Import expenditure can be financed by aid inflows and by borrowing, but in the longer run, a country will find it easier to rely on food imports if it can finance these imports from its own foreign earnings. Foreign earnings include not only merchandise trade but also service export earnings and migrants' remittances (on the debit side, unavoidable debt service might be deducted). Figure 3.4 shows the trends for particular developing country groupings. This chart shows a less reassuring picture for the shorter period 1995-2011. A sharp downward trend in the coverage ratio is evident only for LDCs but with some reversal in recent years. In the case of SIDSs and

LIFDCs, the fall in the coverage ratio was gentler in the earlier years and again there was a deterioration in the coverage ratio during the recent food price spike, particularly for the SIDSs. Nonetheless, for developing countries in aggregate, there is no support in these figures for the view that food import bills are becoming unsustainable. Aggregate figures may, of course, conceal difficulties experienced by particular countries.

Figure 3.4. Ratio of food import expenditure to total export earnings from goods and services

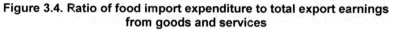

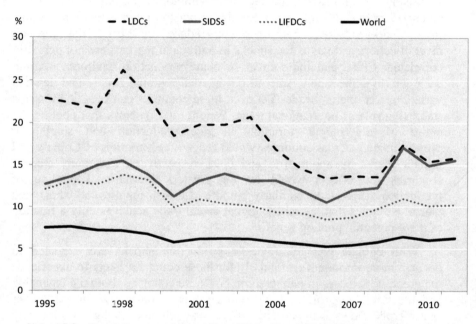

Note: LDCs (Least Developed Countries), SIDCs (Small Island Developing States), LIFDCs (Low-Income Food-Deficit Countries).
Source: UNCTADSTAT.

3.4. Trade and the stability of domestic markets

The spikes in global food prices over the 2007–12 period shook confidence in the stability and reliability of world food markets. In many countries, they led to increased trade policy interventions (Jones and Kwiecinski, 2010; Demeke et al., 2009). In some cases they have re-awakened interest in food self-sufficiency, which, if enforced through trade policy, implies prohibitive levels of protection. The role of stabilisation policies, including the use of stocks and trade policies, is discussed in the study by Abbott (Abbott, 2012) in the OECD study *Agricultural Policies for Poverty Reduction* (OECD, 2012b). That study sets out the basic challenge of managing price risks emanating from both international and domestic markets in those developing countries in which private market solutions may not be possible and social safety nets might not yet be in place.

International trade plays an important role in reducing price risk through enabling countries to make use of world markets in the face of domestic production volatility. The world output of individual food commodities is much less variable than output in individual countries, so greater trade integration holds considerable potential for stabilising food prices. This stabilising role relies on the fact that commodity output shocks across countries are weakly correlated. Similar to portfolio diversification, a move from autarky towards free trade can reduce total price risk through diversification, as long as the shocks in individual markets are not perfectly correlated. China and India cover so many production environments that each can, to some extent, smooth out internal regional supply and demand variations via internal trade. Yet even these countries can benefit from this stabilising role of international trade. Wright (2012) shows that pooling the entire world's output variation in rice production and sharing it proportionately across countries would reduce the variation of China's and India's shares by about 40% and 60%, respectively. For many smaller countries the effects would be far greater. International pooling of production risks could similarly smooth national supplies of wheat and maize. He notes that, currently, global cereal trade achieves only a fraction of these potential pooling benefits.

With climate change, domestic production shocks are expected to become more important particularly for those countries likely to experience the greatest increases in temperature. For these countries, both the balancing and the stabilising role of trade will become increasingly important over time. Trade flows can partially offset local climate change productivity effects, allowing regions of the world with positive (or less negative) effects to supply those with more negative effects. This stabilising role is illustrated by a simulation of an extended drought in South Asia, which begins in 2030 with a return to normal precipitation in 2040 (Nelson et al., 2009). The analysis shows how substantial increases in trade flows could soften the blow to Indian consumers. Large increases in imports (or reductions in net exports) of rice, wheat, and maize result in higher world market prices, implying that other countries' producers and consumers help to reduce, though certainly not eliminate, the suffering that a South Asian drought would cause.

A direct comparison of production variability at global and national levels assumes that all global production is potentially available to meet a national shortfall. In practice, the share of global production of the major staples entering international trade is rather low (Figure 3.5). Agricultural commodity markets are described as thin markets, meaning that relatively small shares of production are traded internationally (Liapis, 2012). When an unusual event takes place, such as the US drought in the summer of 2012, the sharp reduction in production is translated into an even sharper fall in exports or increase in imports, putting immense pressure on markets where only a fraction of production is traded internationally. This can lead to sharp volatility in the prices of agricultural commodities, as witnessed in 2007/8 and 2010/11, particularly if global stocks are low. For example, a 2% decline in milled rice production (9.2 million tonnes in 2010) equates to 28% of world trade in rice in 2010. The impact of market thinness on volatility may be magnified if, in addition, there is a high concentration of export suppliers. Rice, for example, is not only a thinly traded product with less than 7% of global production entering the world market, but trade is also highly concentrated with only six countries accounting for 90% of global rice exports.

Figure 3.5. Proportion of global grain production traded globally

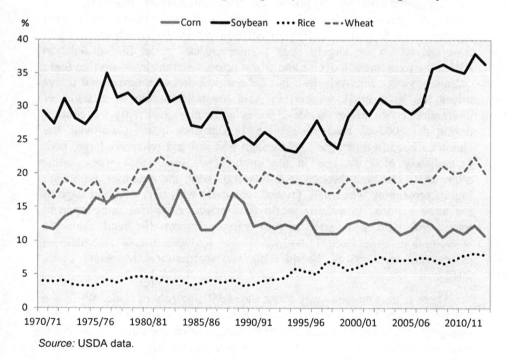

Source: USDA data.

The reason why world agricultural markets are thin is because national boundaries impede the full transmission of world to domestic prices and vice versa. If there was perfect and instantaneous price transmission, then the world market would encompass global production and consumption and not just that element which is traded between countries. In practice, domestic agricultural markets are far from fully integrated into world markets. There are many reasons for this lack of integration. They include fluctuating exchange rates, high transport and transactions costs leading to significant differences between export and import parity prices, market distortions and price controls set by governments, the persistence of trade barriers, and market imperfections. Of special interest are those barriers to market integration resulting from government policy. These can include high transport costs (often arising from inadequate competition in road transport markets) as well as border interventions deliberately designed to prevent the transmission of world market price instability into domestic markets, such as quotas, variable import levies, export restrictions and similar measures. Limited price transmission exacerbates global price fluctuations, but at the same time may serve to protect domestic agents from the full severity of international price volatility (Keats et al., 2010).

The degree to which prices are transmitted from international to domestic markets varies widely among regions. Among developing countries, the largest pass-through is observed in the countries of Latin America, which are largely open to international trade. In Sub-Saharan Africa the pass-through of rice and wheat prices to countries importing these cereals has been relatively fast, but the transmission of international maize prices has been much weaker. In Asia the transmission of changes of international rice prices to local prices differed significantly by country during the 2007-08 food price spike. In Bangladesh and Cambodia, the countries open to trade, the pass-through was fast and relatively large, both immediately after the rise in the international price and three months afterwards. The pass-through in China and India, the countries with high import protection, was small. Overall, countries with high net food imports are more exposed to volatile world food prices. Countries more open to trade, and with a larger share of cereals imports in total domestic consumption, experience faster and larger transmission of international prices onto local prices (World Bank and International Monetary Fund, 2012).

There is thus an ambiguity about the stabilising role of trade. While the portfolio diversification effect contributes to price stabilisation, countries engaging in trade also run the risk of importing price instability. This risk is amplified when markets are thin. However, increasing staple food self-sufficiency to reduce dependence on the world market would not necessarily eliminate food price volatility. While it would decrease volatility due to international markets it would increase volatility due to domestic supply shocks. Thus, in assessing the stabilising role of trade from the point of view of developing countries, the appropriate comparison is between the variability of domestic prices due to domestic shocks to supply and demand and the variability due to global prices.

Abbott (2012) notes that on balance, domestic shocks are more frequent and more severe than international shocks, yet that large international price spikes recur periodically. Other evidence suggests that, even allowing for the current imperfect nature of world markets, the stabilising role of trade is the dominant influence. Ivanic et al. (2011) compare the levels of domestic price volatility for four major crops (maize, rice, soybeans and wheat) under two scenarios: one assuming the current level of trade and protection and one in which international trade in these commodities is abolished.[7] Their results show that international trade – with very few exceptions – lowers domestic price volatility, in many cases very significantly; for example, the standard deviation of rice and soybean prices in East Asia drops from nearly 30 percentage points to less than five percentage points. The introduction of trade with no policy interventions helps greatly lower domestic price volatility by allowing those regions with better harvests to supply output to those regions with worse harvests. This stabilising capacity of international trade is possible because crop yields are only very weakly correlated across regions, which means that simultaneous global crop failure is extremely unlikely.

Minot (2011) quotes several pieces of evidence to suggest that, in the case of Sub-Saharan Africa, price volatility due to domestic supply shocks is as large as or larger than volatility due to international markets. For example, the price of maize in South Africa (which is a source of imported maize for its neighbours) is more stable than the price of maize in most other Sub-Saharan African countries, and the estimated import parity price of US maize in Sub-Saharan Africa is more stable than the domestic price of maize in most of these countries. More generally, the price volatility of internationally tradable products is lower in Sub-Saharan Africa than that of non-tradable commodities and commodities that are tradable only on regional markets (World Bank and International Monetary Fund, 2012). For example, wheat, rice, and cooking oil — products that are imported on the African continent — exhibit lower price volatility than the prices of domestically produced staples. Efforts to increase the tradability of these less traded commodities would help to lower their high domestic price volatility.

Developing countries worry not only about importing price instability from world markets but also about the possibility that sudden sharp increases in import volumes can disrupt their domestic markets. There has been extensive investigation of the importance of such import surges in recent years (Sharma, 2005). While definitions of what constitutes an import surge differ, it is clear that, as a statistical phenomenon, import surges are very frequent. However, while the incidence of surges may have risen, and surges appear to be a fairly common phenomenon in developing countries, these figures tell us nothing about the impact of the surges. There is nothing either inherently "good" or "bad" about an import surge. Rising imports are not necessarily a negative thing for developing countries, as they add to food availability and to the reduction of hunger. It is often presumed that an import surge of a particular commodity disrupts local markets and pushes down prices, negatively affecting the livelihoods of people relying on the production of that commodity. De Nigris (2005) examined correlations between import surges (measured in per caput terms) and production per caput. He found many examples of negative correlations, indicating an inverse relationship between imports and domestic production and suggesting that imports were needed to compensate for domestic shortfalls. He also found positive correlations for other products where imports increased at the same time as domestic production and which probably reflected increasing demand for these products generated by economic growth. Sharma (2005) also found many cases where an import surge occurred even while domestic prices continued to rise, leading him to conclude that imports have been 'pulled in' through prior shortfalls in domestic production rather than higher imports causing domestic production to fall. Thus, the consequences of increased imports for food security need to be carefully evaluated before deciding on the appropriate response.

Since the price spikes in the 2007-12 period, more attention has been paid to the consequences of world price instability for food security in developing countries than to the consequences of import surges. During this period, many countries pursued trade and domestic policy responses intended to stabilise domestic markets and protect urban consumers (Abbott, 2009; Jones and Kwiecinski, 2010). A number of key grain exporting countries, primarily developing economies, adopted export bans or at least partial export restrictions in an attempt to provide enough domestic production for local consumption. At the same time, some major grain importing nations reacted by tendering larger-than-anticipated import bids, reducing pre-existing import restrictions such as tariffs and relaxing tariff rate quotas.

The use of trade measures to insulate economies from shocks to world prices can, at best, transfer the risks associated with commodity production and trade. If many countries seek to transfer price risk to others, the outcome is likely to be ineffective (Martin and Anderson, 2012). In the case of a large exporter, or if a number of exporting countries that are collectively large in the market impose export restrictions, the effect is to increase the world price of the staple food. This increase offsets some of the impact of the reduction in the domestic price. If, in addition, importing countries reduce tariffs on food imports in an attempt to avoid adverse impacts on consumers, the increases in world prices resulting from the initial price shock and the restrictions imposed by exporters will be further compounded. Thus, the attempts by exporters and importers to offset the impacts of a price increase on themselves may be self-defeating. If all countries follow this type of policy, the stabilising impact on domestic prices is, on average, eliminated, although countries that insulate more than others may experience reductions in price volatility, while those who insulate less may experience increases in price volatility.

Despite declared intentions, government interventions are not always successful in stabilising domestic market prices. Anderson and Nelgen (2012) compare the variability of domestic prices relative to border prices for various developing country regions and for high-income countries for the periods 1955-84 and 1985–2004 (that is, before and following the major economic policy reforms that began for many countries in the mid-1980s). Among developing country regions, the ratios are between two-thirds and four-fifths for Asia, quite close to one for Latin America, and close to or slightly above one for Africa. Interventions in developing Asia are thus shown to be somewhat effective in providing insulation against world market volatility. Asian rice producing and consuming countries have a long history of using border measures to successfully stabilise domestic prices (Timmer, 2010). In contrast, interventions in Africa were such as to possibly even destabilise domestic markets. Taken together, the indicators for the world as a whole suggest that market interventions by governments appear to have had very little impact in preventing domestic market prices from gyrating less than prices in international markets (Anderson and Nelgen, 2012).

IFPRI research has shown that these trade restrictions can explain as much as 30% of the increase in prices in the first six months of 2008 (von Grebmer et al., 2011). Yu et al. (2011) find that the trade policy responses in various countries had differential impacts on the prices of agricultural commodities. Their simulation results show that the overall world price impact of trade policy distortions is most significant for rice, at 24%, followed by wheat (14%) and barley (9%). Poorer food-deficit countries and regions, with limited power to manipulate their trade policies, experienced higher price increases than those major trading countries which adopted policy interventions. The authors show that developing countries which are net importers but did not implement trade policy interventions experienced significant welfare losses resulting from interventions implemented by other major trading countries.

In addition, by lowering domestic prices, export restrictions reduce the incentives to increase production and for those who can – in particular the relatively well-off members of the community – to reduce their consumption. The lower prices penalise farmers, reducing the incentives for investments that can increase long-term supply. While an export restriction (but not an export ban) can improve an exporting country's terms of trade and thus possibly its overall economic welfare, in general there are almost always more efficient instruments than trade measures to avert losses for politically significant interest groups (Anderson and Nelgen, 2012). Small and vulnerable developing countries may not be able to avail of insurance against price volatility or make use of direct measures to target poor households (in periods of high prices) or affected producers (in periods of low prices). In these circumstances, trade measures can be shown to be second-best complements to storage policies (Gouel and Jean, 2012). However, such trade interventions are not a co-operative way to address price volatility and can actually exacerbate it. If trade measures are unavoidable, the challenge is to design agreed rules which can limit the negative spillover effects on other countries.

In assessing the need or otherwise for market stabilisation, it is important to identify the origin of risk (international or domestic), the degree of exposure, and the nature of the consequences. The role for national policies will depend partly on the extent to which price volatility is contained at the international level. World price changes may be transmitted onto domestic markets more fully in some countries than others, and more in some years than in others. Domestic shocks, stemming chiefly from production shortfalls, are typically more frequent than international shocks, so market openness may help reduce the frequency of shocks. But such a policy may not be sufficient to contain rare but severe international shocks. The worst case scenario is one where domestic and international shocks reinforce each other, for example when the domestic harvest fails and the government needs to purchase large amounts of imports, and there is a price spike on the world market. The priority under these circumstances is to ensure that poor countries are provided with the instruments to address this rare but potentially disastrous scenario.

Governments may also be able to use market based mechanisms to help mitigate shocks that can affect the balance of payments and lessen their ability to implement social programmes. For example, Malawi has implemented a subsidised weather-indexed insurance programme which helps to finance food imports when weather related domestic production shortfalls occur. Governments may also use option contracts to lock in future food import purchases, so that future import costs are known in advance. Increased international assistance – financial and technical – may be required to help put these mechanisms in place.

Price volatility is one element of a wide variety of risks to food availability that governments seek to manage. These include risks emanating from the world market, such as price spikes, embargoes and a breakdown of trade, as well as risks with domestic origins, such as wars, civil conflict and drought. The optimal approach to managing these risks would consist of parameterising them (i.e. working out the distribution of probabilities), understanding their interactions, and choosing the mix of policy instruments that would minimise a "loss function" – such as the number of people affected by food insecurity. Those impacts would need to be related to a quantifiable indicator.

Different instruments are suitable for managing different types of risk. But those instruments also have interactions. For example, if effective systems of social protection are in place, that may limit the need for other measures to contain the impacts of a food price spike on consumers.

Not all risks can be parameterised, but the principle of undertaking a rigorous assessment of risks and linking those risks to specific food security outcomes can be followed. Current OECD work is employing such an approach, with a view to suggesting policy portfolios that can respond to risks in a manner that reflects their relative probabilities and potential impacts on food security. The principles of such an approach are outlined in Box 3.2.

Box 3.2. A portfolio approach to risk management

Food may be available and accessible today, but the risk of it becoming unavailable or inaccessible in the near future may persist. Several unforeseen events can put food security at risk.

Households can suffer from income losses and health problems, and be exposed to sudden increases in food prices. They can also be afflicted by humanitarian crises due to conflicts or extreme natural events. Farm households and agricultural producers can experience production shortfalls due to bad weather conditions such as insufficient rain or too much rain in the wrong season; extreme temperatures; or animal or plant diseases. They can also suffer from sudden reductions agricultural prices if they are net sellers of food staples, or sudden increases if they are net buyers.

At the aggregate country level these individual risks can become a risk for the whole society. This happens when the sources of risk are systemic, that is, they affect significant geographical areas and social groups, and they are large enough to surpass a socially tolerable threshold. This can occur when a drought affects an area producing a significant share of domestic food production, threatening regional or national food security. It could also result from floods or

---

exposure to contagious pests or diseases. Sudden price hikes, trade interruptions or embargos, and increases in balance of payments deficits can hinder the country's capacity to import food. Humanitarian emergency situations – due to natural disasters, conflicts or accidents – can also pose national level risks to food security.

Individuals, societies and governments have legitimate concerns about such eventualities and need to take decisions to protect against the risk of becoming food insecure. Risk management tools and strategies can help to manage these risks, and accountable governments need to ensure that the responsibilities of managing different layers of food insecurity risks are efficiently distributed between individuals, local communities, market instruments and government policies. At the same time the government has to ensure that the appropriate instruments and tools are accessible so that all agents can develop efficient risk management strategies.

As a rule of thumb, governments have two roles to play (OECD, 2011): first, creating information and knowledge of risk management techniques to facilitate risk assessment and the development of risk management markets; second, developing policy instruments to manage systemic and catastrophic risks that are beyond the coping capacities of individuals and communities. The first aspect of this role facilitates the management of risk at individual household level and is discussed in Chapter 4 on access to food. The second aspect refers to the management of the country level risk of food insecurity. Both household and country level risk management in relation with food security are the focus of international initiatives such as the G20 supported Platform on Agricultural Risk Management (PARM).

Beyond policies to improve the level of food security, governments also implement policies to ensure stability or resilience to shocks. Most societies demand that their governments ensure some minimum level of food security, reflecting what the country can afford. The country level stability objective requires managing different threats to this minimum level of food security. Policy choices must be able to respond to different sources of risk, from high world market price volatility to domestic crop failures. Potential policy responses include fostering a competitive domestic food sector and ensuring a stable physical and financial access to imports.

Following good risk management practices, managing food security risk at the country level should begin with a rigorous assessment of the risks that threaten the country's food security. Risk assessment starts from collecting stakeholders' perceptions of food security risks, and experts' assessments of potential threats to the prevalence of food security. This information will identify specific events or plausible scenarios that could potentially reduce the prevalence of food security below an acceptable level in that country, including those with low probability but severe consequences. Given much uncertainty and potential misperceptions about the nature and probability of threats, experts and stakeholders need to work together to identify credible risks. The food insecurity risk scenarios should assess the approximate likelihood of a given scenario and estimate its likely consequences for food security.

Once the implications of different food insecurity scenarios are assessed, the optimal risk management policy will consist of a portfolio of responses that is able to tackle food security risks from different sources. Such an approach will typically lead to a diversified portfolio of risk management instruments. Fears about trade disruptions, catastrophic droughts, food price spikes and other food insecurity risks can trigger different policy responses, ranging from the provision of social safety nets to the pursuit of food self sufficiency. But the response to one type of risk could be ineffective or even be counter-productive with respect to another. The best policy package will take account of the probability and consequences of different scenarios; how different instruments deal with not just the type of risk they are designed to counter, but also how they affect the outcomes of other risks, and finally their implications for other non risk related objectives. Such an approach to risk assessment can facilitate more rational responses to the fears of food insecurity. Ongoing work of the OECD on Indonesia and, possibly, on other emerging economies will help to test and develop methods for food security risk assessment at government level.

Source: OECD (2012c).

Notes

1. LDCs are defined by the United Nations, LIFDCs are defined by FAO and NFIDCs are defined in the WTO Agreement on Agriculture.

2. The definition of food adopted in this study includes those agricultural commodities that are considered to be most important for basic diets: cereals, meat, dairy and eggs, vegetable oils, and sugar

3. McCalla and Valdés (1999, 2004) were the first to construct a taxonomy of developing countries and use this to look at specific interests of developing countries in trade reforms. This taxonomy based on net trade position and income category has been updated in Valdés and Foster (2012). They distinguish between net agricultural exporters/importers and net food exporters/importers, defining food narrowly as a subset of staple commodities. They reach the same conclusions as in the text that there has been a transition from net agricultural exporters to net agricultural importers.

4. Another issue is that food imports include not only staple foods but also some highly-processed foodstuffs and alcoholic beverages which make only a limited contribution to food security. Valdés and Foster (2012) in their analysis restrict the definition of food to staple foods only.

5. A regression of agricultural growth on the growth in food imports shows a negative relationship, but one that is very weak, with an R^2 value of only 0.01 (Matthews, 2012).

6. He also calculates that the share of the cost of aggregate food imports to the aggregate merchandize exports of LDCs averaged some 60% over the same period, with a very wide spread for individual countries. (This figure is much higher than the figure shown below based on total foreign exchange earnings from UNCTADSTAT and the differences need further investigation.) The NFIDCs' share amounted to an overall simple average of cost of food imports to merchandize export earnings of about 12%, with individual shares ranging from 3% to just over 100% and the maximum for any country not exceeding 115% at any year during the 1990-2009 period.

7. The authors use the standard GTAP model with its stochastic extension to calculate the covariance of global and domestic prices as a result of the exogenous covariance matrix of regional yields for maize, rice, soybeans and wheat.

References

Abbott, P. (2012), "Stabilisation policies in developing countries after the 2007-08 food crisis", in *Agricultural Policies for Poverty Reduction*, OECD Publishing, Paris.

Abbott, P. (2009), "Development dimensions of high food prices", *OECD Food, Agriculture and Fisheries Papers* No. 18, OECD Publishing, Paris.

Alexandratos, N. and J. Bruinsma (2012), "World agriculture towards 2030/50: The 2012 revision", *ESA Working Paper* No. 12-03, FAO, Rome.

Anderson, K. and S. Nelgen (2012), "Trade barrier volatility and agricultural price stabilization", *World Development* 40 (1), pp. 36–48.

de Nigris, M. (2005), "Defining and quantifying the extent of import surges: Data and methodologies", *FAO Import Surge Project Working Paper* 2, FAO, Rome.

Demeke, M., G. Pangrazio and M. Maetz (2009), "Country responses to the food security crisis: Nature and preliminary implications of the policies pursued", *Initiative on Soaring Food Prices*, FAO, Rome.

Dorosh, P.A., S. Dradri and S. Haggblade (2009), "Regional trade, government policy and food security: Recent evidence from Zambia", *Food Policy* 34, pp. 350–366.

FAO (2012a), *The State of Food Insecurity in the World*, FAO, Rome.

FAO (2012b), *State of Food and Agriculture. Investment in Agricultural for Food*

FAOSTAT (2012), *FAO Statistical Database*, FAO, Rome.

Foresight (2011), *The Future of Food and Farming: Challenges and Choices for Global Sustainability*, The Government Office for Science, London.

Fulponi, L., M. Shearer and J. Almeida (2011), "Regional trade agreements - Treatment of agriculture", *OECD Food, Agriculture and Fisheries Working Papers*, No. 44,

Global Harvest Initiative (2012) The 2012 Global Agricultural Productivity (GAP) Report, Washington DC.

Gouel, C. and S. Jean. (2012), "Optimal food price stabilization in a small open developing country", *Policy Research Working Paper* No. 5943, The World Bank, Washington, DC.

von Grebmer, K., et al. (2011), "The challenge of hunger: Taming price spikes and excessive food price volatility", *Global Hunger Index 2011 Report*, International Food Policy Research Institute, Washington, DC.

GTZ (2010), "Regional agricultural trade for economic development and food security in Sub-Saharan Africa. Conceptual background and fields of action for development cooperation", GTZ, Eschborn.

Gustavsson, J., C. Cederberg, U. Sonesson, R. van Otterdijk and A. Meybeck. (2011), "Global Food Losses and Food Waste: Extent, Causes and Prevention", Study conducted for the International Congress *Save Food!*, Düsseldorf, FAO, Rome.

International Sustainability Unit (2011), *What Price Resilience? Towards Sustainable and Secure Food Systems*, ISU, London.

Ivanic, M. and W. Martin (2008), "Implications of higher global food prices for poverty in low-income countries", *Agricultural Economics*, 39(s1), pp. 405-416.

Ivanic, M., W. Martin and A. Mattoo (2011), "Welfare and price impacts of price insulating policies", Presented at the 14th Annual Conference on Global Economic Analysis, Venice, Italy, *GTAP Resource* No. 3651, West Lafayette.

Jones, D. and A. Kwiecinski (2010), "Policy responses in emerging economies to international agricultural commodity price surges", *OECD Food, Agriculture and Fisheries Working Papers* No. 34, OECD Publishing, Paris.

Keats, S., S. Wiggins, J. Compton and M. Vigneri (2010), "Food price transmission: Rising international cereals prices and domestic markets", *Project Briefing* 48, Overseas Development Institute, London,

Koester,U. (1986), "Regional cooperation to improve food security in southern and eastern African countries", *Research Report* No. 53, International Food Policy Research Institute, Washington, DC.

Konandreas, P. (2012), "Trade policy responses to food price volatility in poor net food-importing countries", ICTSD Programme on Agricultural Trade and Sustainable Development, *Issue Paper* No. 42, International Centre for Sustainable Trade and Development, Geneva.

Kummu, M., H. de Moel, M. Porkka, S. Siebert, O. Varis and P.J. Ward (2012), "Lost food, wasted resources: Global food supply chain losses and their impacts on freshwater, cropland, and fertiliser use", *Science of the Total Environment*, 438, pp. 477–489.

Liapis, P. (2012), "Structural change in commodity markets: Have agricultural markets become thinner?", *OECD Food, Agriculture and Fisheries Papers* No. 54, OECD Publishing, Paris.

Martin, W. and K. Anderson (2012), "Export restrictions and price insulation during commodity price booms", *American Journal of Agricultural Economics*, 94 (2), pp. 422–427.

Matthews, A. (2012a), "Global food security and the challenges for agricultural research", Presentation to the Conference 'Innovation and competitiveness of the agrarian sector of the EU', Prague, 17 September.

Matthews, A. (2012b), "Agricultural trade and food security", Background paper prepared for OECD.

Mellor, JW. (1966), *The economics of agricultural development*, Cornell University Press, USA.

Mellor, J.W. and B.F. Johnston (1984), "The World Food Equation: Interrelations Among Development, Employment, and Food Consumption", *Journal of Economic Literature*, 22, (2), pp. 531-574.

Minot, N. (2011), "Transmission of world food price changes to markets in Sub-Saharan Africa", *IFPRI Discussion Papers* No. 1059,IFPRI, Washington, DC.

Nelson, G., A. Palazzo, C. Ringler, T. Sulser and M. Batka. (2009), "The role of international trade in climate change adaptation", *ICTSD-IPC Platform on Climate Change, Agriculture and Trade Series Issue Brief* No. 4, ICTSD, Geneva.

Nelson, G., M. Rosegrant, A. Palazzo, I. Gray, C. Ingersoll, R. Robertson, S. Tokgoz, T. Zhu, T.Sulser, C. Ringler, S. Msangi, and L. You (2010), "Food security, farming, and climate change to 2050: Scenarios, results, policy options", *IFPRI Issue Brief* No. 66, IFPRI, Washington, DC.

OECD (2012a), *Policy Framework for Investment in Agriculture*, OECD Publishing Paris.

OECD (2012b), "Policy coherence and Food Security: The effects of OECD countries' agricultural policies", Paper prepared for OECD Global Forum on Agriculture, 26 November 2012, Paris.

OECD (2012c), *OECD Review of Agricultural Policies: Indonesia 2012*, OECD Publishing, Paris.

OECD (2011), *Agricultural Policy Monitoring and Evaluation: OECD Countries and Emerging Economies*, OECD Publishing, Paris

OECD (2006), *Agricultural Policy and Trade Reform: Potential Effects at Global, National and Household Levels*, OECD Publishing, Paris.

Romer Løvendal, C. and M. Knowles (2005), "Tomorrow's hunger: A framework for

Sen, A. (1980), "Famines", *World Development* 8(9), pp. 613-621.

Sharma, R. (2011), "Review of changes in domestic cereal prices during the global price spikes", *Agricultural Market Information System*, Rome.

Sharma, R. (2005), "Overview of reported cases of import surges from the standpoint of analytical content", *FAO Import Surge Project Working Paper* No. 1, FAO, Rome.

SWAC (Sahel and West Africa Club to go in the glossary) (2012), *Settlement, Market and Food Security*, OECD Publishing, Paris

Swinnen, J.F.M. (2010), "The Right Price of Food", LICOS Discussion Paper 259/2010, Leuven, Belgium.

Swinnen, J.F.M. (2007), *Global Supply Chains, Standards and the Poor*, CABI publications, Oxford.

Timmer, P. (2010), "Management of rice reserve stocks in Asia: Analytical issues and country experience", in *Commodity Market Review* 2009-10, pp. 87–120, FAO, Rome.

Timmer, P.C. (1998), "The agricultural transformation", in C.K. Eicher and J.M. Staatz (eds.), *International Agricultural Development*, Johns Hopkins University Press, Baltimore.

Valdés, A. and W. Foster (2012), "Net food-importing developing countries", *Issue Paper* No. 43, ICTSD, Geneva.

Valdés, A. and W. Foster (2007), "The breadth of policy reforms and the potential gains from agricultural trade liberalization: An ex post look at three Latin American countries", in McCalla, A.F. and J. Nash (eds.), *Reforming Agricultural Trade for Developing Countries, Volume One: Key Issues for a Pro-poor Development Outcome of the Doha Round*, pp. 244-296.

Valdés, A. and A.F. McCalla, (1999), "Issues, interests and options of developing countries", Paper prepared for World Bank, Washington, DC.

Valdés, A. and A.F. McCalla, (2004), "Where the interests of developing countries converge and diverge", in *Agriculture and the New Trade Agenda: Creating a Global Trading Environment*, M. Ingco and L.A. Winters, editors, Cambridge University Press, UK.

World Bank (2012b), "Africa can help feed Africa: Removing barriers to regional trade in food staples", *Africa Region Report* No. 66500-AFR, The World Bank, Washington, DC.

World Bank (2007), *World Development Report 2008: Agriculture for Development*, The World Bank, Washington, DC.

World Bank and International Monetary Fund (2012), *Global Monitoring Report 2012: Food Prices, Nutrition, and the Millennium Development Goals*, The World Bank, Washington DC.

Wright, B. D. (2012), "International grain reserves and other instruments to address volatility in grain markets", *The World Bank Research Observer*, 27 (2), pp. 222–260.

Yu, B. and A. Nin Pratt (2011), "Agricultural productivity and policies in Sub-Saharan Africa", *IFPRI Discussion Paper* 01150, International Food Policy Research Institute, Washington, DC.

Chapter 4

Improving access to food

This chapter focuses on the ways in which agricultural and broad-based rural development can contribute to improvements in food security. It examines the ways in which governments can strengthen agricultural incomes, while enabling households to diversify their income sources and take advantage of non-farm employment opportunities.

4.1. Lack of access is the main obstacle to food security

The foremost cause of food insecurity is a lack of access, which stems from people not having what Sen referred to as the "entitlements" necessary to provide them with adequate food and nutrition (Sen, 1984). Those entitlements can derive from production (growing food), trade (buying food), own-labour (working for food), and inheritance and transfers (being given food). The foremost reason for households lacking access is poverty and deficient incomes.

The World Bank's poverty measure, which is based on consumption surveys, reflects the notion of food access because it attaches a monetary value to food consumption from each of these four origins. The extreme poverty threshold of USD 1.25-per day is the mean of the national poverty lines in the world's poorest 15 countries, while the USD 2-per-day level is the median of the national poverty lines for developing countries as a whole. Those national poverty lines are established on the basis of a "cost of basic needs" calculation, which contains a food component specific to each country that yields a stipulated food energy requirement. Poorer countries tend to set the parameters of the national poverty line lower (Chen and Ravallion, 2010).

At the global level, the World Bank calculates that there are about 1.3 billion people living on less than USD 1.25 a day. At the same time, the FAO estimates that there are about 850 million undernourished. The FAO calculation is based on an assessment of national food availability and an estimate of its distribution across the population, with the number of undernourished corresponding to those whose consumption falls below a minimum level of calories. Methodological differences, and different yardsticks for minimum energy needs, mean that the World Bank and FAO estimates are not strictly comparable. However, it would appear that some households are extremely poor yet not demonstrably undernourished.

Yet access to food is a pre-requisite for food security, so the 1.3 billion figure can still be interpreted as a low estimate of the number of food insecure. People with an income of between USD 1.25 and USD 2 per day – a further 1.2 billion – are also poor. They may be able to afford an adequate intake at a given moment in time, but the "stability" of their access may not be guaranteed in the face of adverse shocks, such as a family member not being able to work because of ill health, or (in the case of farmers) a poor harvest. By this reasoning, over one-third of the world's population is effectively too poor to enjoy full food security.

At the same time, there are people who are not poor yet suffer from poor nutrition – if not in terms of basic calorie consumption, then in terms of the diversity of diet and other determinants of nutrition that are necessary to ensure an active and healthy life. These "utilisation" aspects of food insecurity are taken up in Chapter 5.

There has been widespread progress on poverty reduction, and the developing world as a whole is on target to meet the MDG target of halving between 1990 and 2015 the proportion of people living on less than a dollar a day. Yet the rate of progress has been uneven, and poverty rates remain high in South Asia and Sub-Saharan Africa, with a similar proportion of people, about 70%, living at less than USD 2 per day in both regions (Figure 4.1). However, there is an important difference between the two continents. South Asia has made much greater progress in reducing extreme USD 1.25 poverty. This suggests that there may be difference in the nature of the food security problem – with a greater proportion of people chronically lacking access in Africa, but a larger share of people managing to afford food yet still vulnerable in South Asia.[1] This may have implications in terms of which policies are needed to strengthen households' incomes and improve their resilience to shocks.

Figure 4.1. Incidence of income poverty by region, 1999 and 2008, %

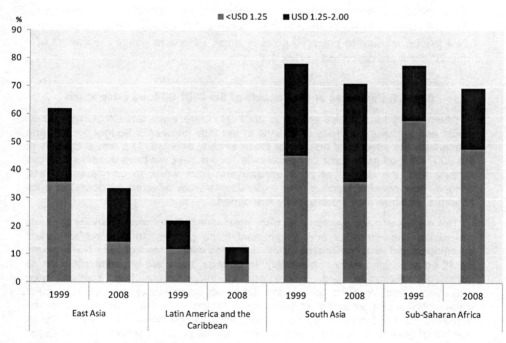

Source: World Bank.

4.2. The importance of incomes versus prices

In low income countries, food consumption expenditures typically account for 50% or more of households' budgets. In lower middle income countries, such as India, the figure is about 40%. Moreover, for the poorest consumer households which do not grow food, food expenditures typically dominate cash outlays. The price of food is thus a key determinant of the household's real income.

Farmers are affected by food prices as both buyers and sellers. The 2008 World Development Report reports that a majority of the rural poor are net buyers of tradable staples (World Bank, 2007), while OECD analysis of nine countries in the World Bank and FAO's Rural Income Generating Activities (RIGA) database finds that the proportion of rural net sellers of food staples is lower than that of rural net buyers in all but one – the exception being Viet Nam. Thus, the effects of food price changes on rural populations are far from clear cut. Furthermore, since rural populations are on average poorer than urban ones (IFAD, 2010) and the poor tend to spend a high proportion of their income on food (reflecting "Engel's law"), they are naturally the most likely to experience hardship in the face of rising food prices. It was these patterns that led to concerns about the impacts of high food prices. Moreover, even short episodes of income loss can cause households to sell productive assets – land and livestock, for example – at low prices, leading to potential poverty traps. Efforts to gauge the impact of the 2007-08 are described in Box 4.1.

Box 4.1. Estimates of the impacts of the 2007-08 food price spike

When world food prices spiked in 2007-08, there were fears that high prices would add perhaps hundreds of millions to the total number of hungry people, and exacerbate the severity of hunger for those already affected. The actual impacts of the 2007-08 food price spike on households' food access will have depended on two factors: first, the degree of price transmission from world to domestic markets; second, how peoples incomes and expenditures were affected by domestic price changes, and how they subsequently responded.

The degree of price transmission from international to domestic markets depends on market characteristics and government trade policies. In a world of perfect information and zero transaction costs, any price differences between two markets would be arbitraged away immediately. In practice, there are impediments such as imperfect information and deficiencies in infrastructure that impede price transmission and lengthen the response time over which markets adjust. The use of trade policies can also restrict price transmission from the international to the domestic market (and vice versa in the case of large countries). In 2007-08, a number of governments responded by mitigating the pass-through of price increases to the national level (Demeke et al., 2009; Jones and Kwiecinski, 2010).

There are several methods to estimate price transmission, from simple approaches like the ratio of percentage price changes between two time periods and correlation coefficients, to cointegration techniques and associated error-correction models. The latter methods overcome the problem of spurious correlation and make it possible to specify the long-run relationship between prices, as well as short-run adjustments. Approaches based on the use of price data alone can be used to calculate the observed degree of pass-through between markets; however, a structural model would be needed to discriminate between the different impediments to pass-through such as policy changes and non-policy factors.

Sharma (2011) computes elasticities of price transmission for maize, rice, wheat and wheat flour in a range of developing countries, calculated as the percentage change in the domestic price divided by the contemporaneous percentage change in the world price. The results indicate that the degree of price transmission was at least one-half in all cases and higher in 2007-08 than in 2010-11. The average exceeded 100%, i.e. domestic prices increased by a greater percentage than world prices, for maize and wheat flour in 2007-08. However care needs to be taken in interpreting these numbers. In the first place they do not establish causality – for example domestic prices in isolated markets may spike by more than the change in world prices because of domestic shocks such as drought. Second, the percentage calculation can be misleading. For an exported product, the domestic price would generally be lower than the world market price, while for an import it would be higher. That being the case, perfect transmission of an absolute price change would result in a transmission rate of more than 100% in the case of an export and less than 100% in the case of an import.

Minot (2011) examines the degree of price transmission between world and domestic markets using cointegration techniques. His analysis considers 62 price series from 11 African countries over periods of 5-10 years. Staple food prices rose by 63% between mid-2007 and mid-2008 – about three-quarter of the proportional increase in world prices. However, African food prices displayed a long-term link to world prices in only 13 of the 62 cases, with rice prices more closely linked than maize prices. It is posited that the global food crisis was unusual in increasing African food prices because of the size of the increase and the fact that it coincided with an increase in oil (and hence fuel) prices.

An interesting finding from the Minot study was that food price increases appeared to be greater in landlocked countries than in coastal countries. This is counter-intuitive, in that one would expect coastal markets to be more integrated with world markets. There are three possible explanations. One is that the markets are not integrated and domestic shocks due to factors such as drought exceeded those emanating from the world market. Another is that the simultaneous rise in fuel and other costs affected transaction costs in landlocked countries more. Finally, grain export bans imposed by several African countries may also have exacerbated the price spike in landlocked countries.

The impact of a given domestic price change on food security depends immediately on how those price changes affect households' food expenditures and (in the case of farmers) production and revenues. Beyond the pure incidence of a price change, the final effect depends on how households respond in terms of their production, consumption, time allocation and savings and investment decisions; and then on wider market and economy-wide linkages. A range of studies have sought to capture some or all of these effects.

Several studies estimate the incidence of world price changes on welfare (real incomes) using household data, without going the extra step of calculating the implications for a specific measure of food security (for example, Ivanic and Martin (2008), Robles and Torrero (2010) and Ferreira et al. (2011), Filipski and Covarrubias (2012)). A general finding is that, with a few exceptions, higher food prices harm more households than they benefit, and they harm the poorest most, since poorer consumers spend more of their incomes on food, while poorer farmers are more likely to be net buyers of staples.

While the studies often make a range of assumptions about price transmission from world to domestic markets, the study by Filipski and Covarrubias, undertaken for OECD, considers observed price changes six staple crops using the FAO's GIEWS database and calculates their welfare impacts in nine developing countries. The results show that, since most rural households are net buyers of staples, they stand to lose from higher staple prices in the short run. However, simulations of the 2007-08 food price crisis suggests that the magnitude and timing of the welfare shocks depended heavily on the type of crops produced and consumed by each rural household. A key variable in all these studies is the proportions of net buyers and net sellers of food staples. The evidence is that the former dominate in developing countries, especially among the poor and potentially food insecure, although there are some exceptions, such as Viet Nam (RIGA datasets used in Filipski and Covarrubias, 2012).

Some studies seek to go beyond capturing the incidence of price shocks and use behavioural models, for example CGEs with households embedded, to calculate wider impacts once household and market responses are factored in (for example, Benson et al. 2008 for Uganda; Warr, 2008 for Thailand). These models can capture significant wider impacts. For example agricultural wage labour may gain from a tightening of the labour market, while farmers may see some of their income gains dissipated in the form of higher input or factor costs. However, none of these studies overturns the result that high food prices tend to affect negatively the welfare of the poor in the short to medium term.

Most simulation-based estimates of the impacts of high food prices have however been *ceteris paribus*. They have not allowed for income growth (or price changes for non-food commodities), which was substantial in many developing countries. For example, the economies of China, India, Brazil, Nigeria, and Ethiopia all grew at 6–10% per year during the food crisis years. Jones and Kwiecinski (2010), in a study of ten emerging economies, note that in aggregate terms, strong GDP growth more than compensated for food price increases in 2007-08, but the situation reversed in 2009 when the persistence of high price coincided with a slowdown in GDP growth. Nor have simulations factored in government policy responses, which ranged from changes in trade and storage policies to food distribution, social safety nets and subsidies. Moreover, it is important to bear in mind that households faced with higher food bills adopt coping strategies. For example they may spend less on education or health, or reduce assets in order to buy food. This may not affect undernourishment immediately, but it could undermine long-term food security. Most short to medium term models do not capture these effects. A further issue with simulations of the impact of high food prices derives from their treatment of cross-border price transmission. As apparent from the above discussion, this is often assumed to be complete; if not, an arbitrary assumption is typically made. Only rarely is the degree of price transmission estimated explicitly and then introduced exogenously into a simulation model.

Headey (2013) used a Gallup World Poll (GWP) series of surveys conducted before, in 2005/06, 2007 and 2008 to investigate directly the impact of the 2007-08 price spike on food security. Interviewees were asked whether they had had any difficulty in affording food and whether they had experienced episodes of hunger over the past 12 months, both at the individual level and at the level of the household. Strikingly, his results suggest that global food insecurity went down from 2005/06 to 2007/08 by about as much as it was said to have increased in previous studies published by the World Bank, USDA, and the FAO. There were exceptions, notably in parts of Africa and Latin America and the Caribbean. He finds that these results are robustly explained by food price inflation and per capita economic growth rates.

Headey's results do not discard a potentially significant impact of the food prices on the incomes and welfare of the poorest, nor on their vulnerability to food insecurity or future food security. However, they suggest that early simulation exercises might have over-estimated the short run impacts of the price spike on food security and that a high degree of uncertainty remains about the impact of recent food price increases on the world's poor. They also underline the need for more research into household adaptation behaviours and into the linkages between economic growth and commodity booms.

Logically, a given increase in the real income of a net food consuming household can be achieved either by raising overall nominal income or by lowering the price of food. Yet fundamentally, there is much more scope for improving food security by raising incomes than there is by lowering prices. Even under the most favourable scenarios, it is unlikely that international cereal prices will fall to the all time lows they reached in the early 2000s. Yet when prices were at that level there were still more than 800 million undernourished people in the world (Figure 4.2).

Of course, food prices still matter. In low income and lower middle incomes countries, where a significant share of the population lives in poverty, net buyer households that formerly enjoyed adequate nutrition may suffer from transitory food insecurity as a result of higher food prices. In the case of farm households, there is a particular concern that net buyer households could be forced to sell productive assets to pay for food and be forced to divert resources – for example by taking children out of school – in ways that jeopardise their long term food security. Yet in principle, higher food prices should be an opportunity for farmers who can respond and achieve a marketable surplus.

While lowering equilibrium food prices would be good for overall food security, everything else equal, the specific effects will depend on how that reduction is achieved. Productivity gains push down prices, to the benefit of food consumers, but the effects on farmers are less clear. For commercial farms with net sales, productivity-led cost reductions are offset by price declines to the extent that markets are competitive. However, early adopters of improved technologies and farm practices benefit, insofar as their cost reductions precede the ensuing fall in prices. Subsistence farmers with negligible net sales or purchases are not affected directly by market price changes, but targeted productivity gains can raise their food consumption and possibly enable them to achieve marketable surpluses. For food deficit farmers, lower purchase costs are an immediate benefit, but, unless the benefits of improved productivity are sufficiently large, could deter them from producing food surpluses. For decades, productivity improvements were the main driver of falling food prices, creating opportunities for innovators, but raising concerns for the incomes of farmers whose productivity languished. Hence, there are legitimate concerns about the effects of both low and high prices on farmers.

Figure 4.2. Undernourishment and food prices over the past two decades

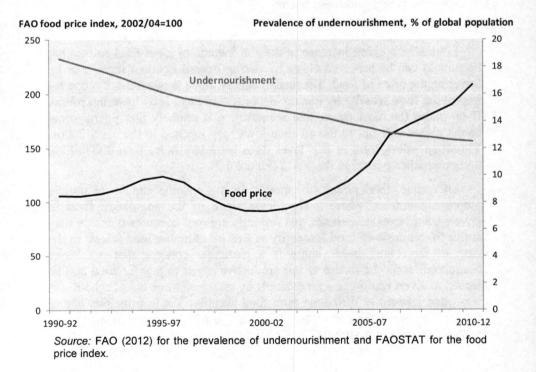

Source: FAO (2012) for the prevalence of undernourishment and FAOSTAT for the food price index.

Rapid income growth can enable consumer households to afford food at a wide range of prices, and the higher income becomes the less the price of basic food staples becomes a significant determinant of their real incomes. As a result, while high food prices may lower living standards in high income countries, for the vast majority of households they do not affect their fundamental food security. For households in upper middle income countries, where per capita incomes are significantly above the World Bank's USD 1.25 and USD 2 per day thresholds, living standards may be significantly reduced by increases in the cost of food. However, food security itself may not be threatened, as the household can economise on other non-food purchases – even items as basic as clothing.

For a given rate of overall growth, a more pro-poor composition of growth allows faster reductions in food insecurity. But some countries, China being a notable example, have grown sufficiently rapidly that poverty has come down despite rising inequality. The key variable is the wages of the poor: if these double or triple, as they have done in China, then that will overwhelm the effects of a similarly rapid increase in food prices.

4.3. Agricultural development as a mechanism for raising incomes

It is beyond the scope of this study to set out the broader requirements for economic growth and poverty reduction. But in many developing countries, the agricultural sector has an important role to play, both as an overall engine of growth, and – because many of the poor work in agriculture or are dependent on agriculture – as a mechanism for pro-poor growth. This section focuses on the particular role of agricultural development.

For most countries, the long-term process of economic development is characterised by a transition from an economy based on agriculture to a more diversified economic structure with larger shares of GDP accruing to manufacturing and services. This process is associated with a parallel reallocation of labour (Figure 4.3).

In middle income developing countries, agriculture's share of GDP is typically 20% or lower, and the sector's role as a basic engine of growth correspondingly less important. In its 2008 *World Development Report*, the World Bank identifies approximately 170 million rural poor living in agriculture-dependent economies (mostly in Sub-Saharan Africa) and a much larger number – 583 million in 2002 – live in transforming economies, a large proportion of them in China and India (World Bank, 2007). The majority of the rural poor in Latin America live in urbanised countries. Globally, about two-thirds of the world's dollar a day poor live in rural areas

(IFAD, 2010). Here, raising rural incomes in general, and agricultural incomes in particular, is essential for poverty reduction.

Smallholder farming is the predominant structure in most rural economies and has therefore received attention as a key vehicle for poverty reduction. Smallholder development can increase returns to assets that the poor possess – their labour and in some cases their land – and if markets are not integrated it can push down the price of staples, which is beneficial when so many of the poor are net buyers of food. Indirectly, the benefits of smallholder growth are likely to be particularly beneficial in agriculture-dependent economies, because of growth linkages to the rest of the economy.

Figure 4.3. Employment shares and variations in per capita income, 2005

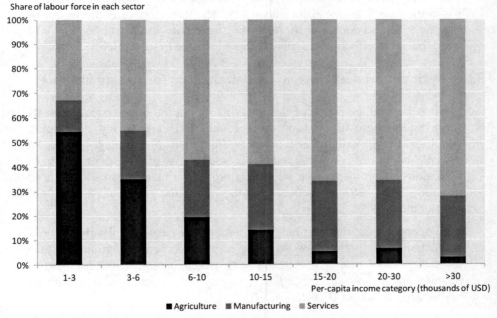

Share of labour force in each sector

Per-capita income category (thousands of USD)

■ Agriculture ■ Manufacturing ▨ Services

Source: World Bank.

The overarching challenge for policy makers is to promote rapid yet enduring poverty reduction. This means achieving income growth and poverty reduction on the basis of current economic structures, while simultaneously facilitating the transformation to economic structures that are fundamentally capable of generating higher incomes. This is not an easy balance to strike, and leads to dilemmas in terms of whether policy makers should focus on raising incomes within agriculture or facilitating the transition out of agriculture; and whether they should be supporting smallholder farming or other – potentially more remunerative – structures.

Agriculture's central role in the early stages of economic development is well documented (Timmer, 1998). Few countries have developed without first developing their agriculture sectors and this remains the key priority in a number of the poorest countries, mostly in Sub-Saharan Africa. Several studies have also suggested that agricultural growth, in particular smallholder-based growth, can deliver more pro-poor development (for example, Hazell et al., 2007), and a number of studies have confirmed that agricultural growth tends to be effective in reducing poverty (Christiaensen et al., 2011; Irz et al., 2001; de Janvry and Sadoulet, 2010).

In poor agriculture-dependent economies, there is therefore a strong case for prioritising agricultural development because of the sector's role as a catalyst for broader economic development. There is also a potential case for re-emphasising agriculture in middle income countries where agriculture is no longer a dominant sector, on the grounds that this would help to close the gap between rural and urban incomes. However, it is also possible to narrow the gap by the absorption of rural labour in the urban economy, and through deeper integration of rural and urban labour markets.

Yet historically agriculture has been discriminated against by developing country policy makers, both in terms of pricing policies (Anderson, 2008) and through urban bias in the allocation of expenditures (Bezemer and Headey, 2008). Until recently, donors have also neglected the sector. Official development assistance to the sector declined in both absolute terms and as a proportion of total allocations, with a fall from USD 8 billion in 1980, equal to 17% of total aid, to a little over USD 3 billion by 2005, corresponding to a share of less than 4%. Recent data show that total official development assistance (ODA) for the wider category of food and nutrition security was around USD 11.7 billion in 2010, up 49% in real terms on 2002. Its share of total ODA stayed fairly constant, at around 7%, with no clear reallocation as a result of the 2007-08 food crisis (OECD, 2012d). The evolution of donor support for food and nutrition security is presented in more detail in Chapter 5.

One reason for the bias against agriculture was low rates of perceived success compared with investments in other areas such as education and health (Easterly, 2008). Another was the combination of declining real agricultural prices and, in successfully developing economies, a falling share of agriculture in GDP and employment. These were interpreted by policy makers as signs of higher returns from investing in other sectors. Yet prices are now high, so that argument is no longer valid, if it ever was[2]. Moreover, the interpretation of agriculture as an inevitably "declining" sector was misplaced. Agricultural investment was and is necessary to elicit the productivity gains that initiate the agricultural transformation which enables resources to be released from the sector and – when part of a balanced development strategy – to be reallocated productively other sectors. The relative "decline" of agriculture is in fact a consequence of development success.

In terms of many of the core fundamentals, there is no need to choose between agricultural and non-agricultural development, because the policies required to foster agricultural development are not sector-specific. The core pre-requisites include the overall investment climate, which depends on factors such as peace and political stability, sound macroeconomic management, developed institutions, property rights and governance. A range of other factors, such as improvements in education and primary healthcare, are not sector-specific either. But rural provision often lags, and greater equality between urban and rural areas would help agricultural development and, just as importantly, promote more balanced rural development. A key feature of these general policies is that they are neutral in terms of the incentives they create: they are good for agricultural development, yet they do not impede people from reallocating time and resources to other activities if those activities are found to be more profitable.

However, there is a case for making investments that are specific to agriculture, in areas such as agricultural research, technology transfer, farm extension and advisory services; and for increasing the share of public spending allocated to sectoral or at least regional public goods, such as rural roads. The gains from investment in agricultural research and development were noted in Chapter 2, while Fan et al. (2008) finds broader evidence that investments in agricultural research, infrastructure and human capital have strong impacts on both agricultural production and poverty reduction. Despite these findings, governments in many developed and developing countries tend to structure agricultural policies more around supporting farmers via price support (and associated trade protection) and income payments, rather than through investments in the sector's enabling environment (OECD, 2011; Anderson, 2008). OECD has developed a policy framework for sustainable investment in agriculture, which is designed as a practical tool to help policy makers create that enabling environment and enhance the development benefits of agricultural investment (Box 4.2).

While there is a need to foster agricultural development, it is important to acknowledge that the exploitation of improved opportunities within the farm sector will itself lead to adjustment stress. For the majority of agriculture-dependent households, their long term (i.e. inter-generational) future lies outside the sector. This is true even in the context of improving opportunities within the agriculture sector as a whole. Hence, while policies need to improve opportunities within agriculture for those who potentially have a competitive future in the sector, they also need to create wider opportunities for those who do not, or could earn higher incomes elsewhere. This underlines the importance of balanced rural development in order to create opportunities outside as well as within agriculture – and ensure that labour is "pulled" rather than "pushed" out of the sector.

Box 4.2. A policy framework for sustainable private investment in agriculture

The Policy Framework for Investment in Agriculture (PFIA) aims to support countries in evaluating and designing policies to mobilise private investment in agriculture for steady economic growth and sustainable development. Attracting private investment in agriculture relies on a wide set of policies that go beyond agricultural policy, including macro-economic and sectoral policies. A coherent policy framework is an essential component of an attractive investment environment for all investors, be they domestic or foreign, small or large. The PFIA is a flexible tool proposing questions for governments' consideration in ten policy areas to be considered by any government interested in creating an attractive environment for investors and in enhancing the development benefits of agricultural investment (Figure 4.4).

Figure 4.4. Policy framework for investment in agriculture

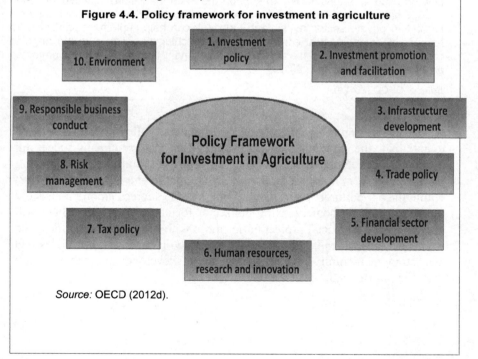

Source: OECD (2012d).

The PFIA draws on the Policy Framework for Investment (PFI) developed at the OECD in 2006 by 60 OECD and non-OECD countries. It was first developed in 2010 by the NEPAD-OECD Africa Investment Initiative, the Sahel and West Africa Club (SWAC) and the Office of the Special Adviser on Africa (OSAA) of the UN Secretary General. It has benefited from inputs from several OECD policy communities, in particular the Secretariats of the Committee for Agriculture, the Development Assistance Committee (DAC), the Committee on Fiscal Affairs and the Committee on Financial Markets.

The PFIA has already been used as a self-assessment tool by Burkina Faso, Indonesia and Tanzania and is currently being used in Myanmar. Given the range and variety of relevant measures involved, the PFIA promotes policy co-ordination at the host-country level, both in the design and implementation phases. All relevant stakeholders including not only Ministries and government bodies but also the private sector, civil society and farmers' organisations, should be actively involved in the PFIA process.

The PFIA can complement existing national and international initiatives aiming to attract more but also better investment in agriculture. In particular, it can contribute to achieving the CAADP and Grow Africa objectives by supporting the design and implementation of regional and national agricultural investment plans and investment blueprints and by strengthening cross-sector collaboration. It can provide the Global Donor Platform on Rural Development (GDPRD) and *the New Alliance for Food Security and Nutrition* with an instrument to facilitate donor dialogue, harmonisation and alignment around countries' priorities. The PFIA can also be used as an instrument to support the Feed the Future initiative launched in 2009 by the US government and aiming to create enabling policy environments facilitating private sector investment. Finally, it can help implement principles for responsible investment at country-level, in particular by building on the ongoing consultations on responsible agricultural investment launched by the Committee on World Food Security in 2012.

Source: OECD (2012a).

So while promoting agricultural development, policy makers also need to anticipate the structural changes in agriculture that accompany successful economic development. This means offering multiple development pathways for farm households: improving competitiveness (i.e. productivity) within the agricultural sector; diversifying income sources among household members; and, for some, leaving the sector for better paid jobs. There will also be a need for social protection for those with limited opportunities, and who have difficulties in adjusting (for example older farmers). Policies that can strengthen rural incomes along these development pathways are discussed in Box 4.3.

Box 4.3. Development pathways for farm and rural households

In the long run, there is a need to anticipate the structural changes in agriculture that accompany successful economic development. These include i) a declining share of agriculture in GDP as the economy develops and diversifies; ii) a release of labour from the sector driven by a combination of the "push" of labour-saving technical change in agriculture and the "pull" of growing labour demand in non-agricultural sectors; and iii) rising agricultural output. This means offering multiple development pathways for farm households: improving competitiveness (i.e. productivity) within the agricultural sector; diversifying income sources among household members; and, for some, leaving the sector for better paid jobs.

Improving the competitiveness of farm households

Given the need to acknowledge that some farmers will succeed while others will not, and the impossibility of identifying exactly which farmers fall into each category, the main role for policy would appear to be in providing public goods that can improve competitiveness, but impose few distortions to incentives at the margin, such as investments in rural infrastructure, skills and training, and R&D.[1] Such investments are unlikely to crowd out the development of other activities and potential income streams, although they are likely to accelerate the pace at which more efficient operators absorb and replace less efficient ones. Most of the relevant expenditures would need to be made at the economy-wide or sectoral level rather than in the form of payments to individuals. A further role for policy is when there are endemic market failures, for example in credit markets. Access to credit is important for smallholders, and private credit markets may find it not worth their while to engage with smallholders, simply because of their size and the difficulties of becoming informed about the creditworthiness of many small operations.

In many developing countries, farmers may have insecure land rights, while land rights rental markets function poorly or do not exist at all. Secure land rights can improve incentives for investment in the land, and can also facilitate the development of rental markets. The latter can in turn help compensate for market failures, provide flexible responses to economic and productive incentives, allow farmers to invest in farming capital, and help the poor and young gain access to land under conditions that are less demanding than those required to participate in land sales markets. Renting land may also be a first step to future land acquisition. The underdevelopment of rental markets may prevent the consolidation of land into more productive units, thus impeding agricultural investment and making it more difficult for uncompetitive farmers to diversify out of the sector.

Income diversification for farm households and salaried agricultural workers

Income diversification is essential for many farm households. For the poorest farm households, this is likely to provide some insurance and is in effect a "coping" strategy. For other farm households, having one or more family members draw income from outside agriculture may be the start of a successful move into more remunerative activities. Policies that support farm income alone, such as market price support, act as a disincentive for income diversification outside agriculture, and create an obstacle to one of the key "adjustment pathways". The key policies required to help households diversify their income sources are again those that improve human capital. Regional development policies, including the development of rural infrastructure, may also have an important role.

Leaving the sector for skilled employment

Ultimately, the majority of smallholders in developing countries, or at least their descendants, will have stronger prospects outside the agricultural sector than within it. The most important need, if not for this generation then for the next, would therefore appear to be investment in the education and skills that would enable households to command higher wages. At the same time well-defined property rights, especially with respect to land, are important for farmers to be able to cash in their assets, and exit the sector on favourable terms.

Regional development programmes, by targeting economic assistance to less developed regions, may also have a role in bringing jobs to people (rather than the other way round) and so can prevent the problems associated with mass migration into cities. However, rural policies are not fundamentally agricultural policies (nor vice versa). Regional policies can boost development within and outside agriculture, but without biasing household decisions about how best to invest for the future.

In many middle income countries the conditions of salaried agricultural work are at least as important as the development of small scale farm entrepreneurs. In Chile, for example, two-thirds of all households receiving the majority of their income from agricultural sources are salaried workers, not farmers. Labour market policies have an important role in ensuring that core standards of employment are met, while improved labour market flexibility has been suggested as a way of reducing informality (OECD, 2008).

Social protection for households that cannot adjust

Many poor households, notably older ones, face severe limitations in their adjustment potential, irrespective of the policies that are in place (for example, resource poor and post retirement age farmers). Hence the need for social protection to address chronic as well as transient income shortfalls. Investments in human capital (notably education) and measures such as contingent cash transfers can ensure that the next generation makes a quantum leap in terms of development.

---

1. There is evidence to suggest that improvements in agricultural productivity have a strong effect in reducing poverty (Irz et al., 2001). There is also evidence that agricultural growth has helped support broader economic growth (for example, Tiffin and Irz, 2006), although agriculture's role as a necessary driver of development has been questioned (Gardner and Tsakok, 2007).

Source: OECD (2012b).

4.4.   The role of smallholders

Smallholder farmers, who dominate agricultural systems in most developing countries, are directly implicated in the agricultural transformation.

Raising their incomes is key to improvements in food security, but there is no clear consensus on the extent to which that should be done by raising their farm incomes directly or by facilitating their capacities to earn incomes outside the sector. Proponents of the former approach (e.g. Hazell et al., 2007; Morris et al., 2009) note that smallholder farming underpins rural economies and forms the basis for the livelihoods of the world's poor, and maintain that the development process has to start by tapping the potential of existing structures. The counter argument is that the improvements in productivity needed to assure higher incomes will require a vast reduction in the proportion of the population engaged in agriculture, and that policy should focus on promoting competitive agriculture and facilitating the exit of non-competitive farmers from the sector (e.g. Collier and Dercon, 2009).

There has been much debate over the relative efficiencies of small versus large farms. A range of benefits from small scale family farming have been noted. For example, farm labour may be easier to motivate and supervise, while smallholders may have important local knowledge and may be more adept at managing some forms of risk. On the other hand, there are important economies of scale beyond the farm in areas such as procuring inputs, obtaining information on markets and technical farming issues, in meeting standards and certifying production, and in transacting with large scale buyers from processors and supermarkets, with their exacting demands (Wiggins, 2009).

However, a fundamental question concerns what types of farm operation are intrinsically capable of generating enough income to ensure the farm household's food security – either directly through production of food or by generating the income from other crops necessary to buy food. That means ensuring not just that annual income is sufficiently high, but also that consumption can withstand shocks, be they idiosyncratic (such as a family member falling ill) or systemic (for example a regional drought). This question might not matter if off-farm income could plug the gap for smallholders. But often there is no such income or it comes from uncertain sources such as casual employment, and the issue comes down to whether it is more cost-effective to raise farm incomes than off-farm incomes.

In some cases, smallholders may lack the assets necessary for effective commercialisation. The most important asset they may lack is land. For example, Jayne et al. (2003) examine the size distribution of land holdings in five countries in southern and eastern Africa – Ethiopia, Kenya, Mozambique, Rwanda and Zambia – and find considerable inequality, with 25% of farm households in all countries having access to less than 0.1 ha if land per capita. Income per capita rises as the size of landholding increases from 0.1 ha to 0.25 ha, and more gradually as the size of landholding increases further. Their overall assessment is that the poor generally lack the land, capital and education to respond quickly to agricultural market opportunities and technical innovation (Jayne et al., 2003, p. 254).

Jayne et al. also make a "best case" assessment of agricultural incomes for representative households in western Kenya. According to their analysis, the 75th percentile farm household has an average of 0.6 ha of land (0.1 ha per capita), albeit land that yields two crops per year. This household could satisfy all its maize requirements at enhanced yields (per capita consumption requirement is about 140 kg per person per annum) and devote 80% of its land area to crops other than maize during the short rains season. However, its income per capita from farming activities alone would still only be around half the international poverty line of USD 1.25 per day. Meanwhile, 25th percentile farm household would not satisfy its maize requirements. The authors conclude that the larger amongst small farms will be first to grow crops for sale and that will have the most realistic opportunities for commercialisation.

If only a subset of smallholders is capable of commercialisation, and of being competitive with other (larger) farm structures, then a short-term choice concerns how much policy effort to invest in raising the farm incomes of the remaining farmers – as opposed to providing social protection and increasing their ability to earn income from other sources. The attraction here is that the application of well-known technologies would appear to be a way of raising incomes rapidly and possibly at low cost. Moreover, enabling a wider group of smallholder to produce more of their own food should improve their food security and improve their prospects of moving onto more rewarding employment in the rural non-farm economy or elsewhere. Yet ultimately, there is an inter-temporal trade-off, and it is important that policies do not impede the transition to structures that are capable of generating higher incomes, or deter people from seizing opportunities outside the sector.

Furthermore, adjustment is rarely seamless. An organic consolidation could involve the smallest land holdings grouping to achieve economies of scale in areas such as selling, procuring inputs, obtaining technical advice and securing loans. Savings on farm management could enable households to retain secure tenure of their land yet still have increased scope to exploit opportunities outside the sector. However, obstacles to smoother transition include uncertainties over land tenure rights in the event of a family stopping to farm, and the difficulties of monitoring non-family labour. As a result, many countries have dualist farm structures with large commercial farms co-existing alongside smallholder (often subsistence and semi-subsistence) structures. The emerging economies covered by the OECD's regular monitoring and evaluation of agricultural policies illustrate this point (Box 4.4). In many of these cases, the smallholder group represents an outstanding social adjustment problem that economic development has so far failed to address.

In terms of the agricultural transformation, we have a good sense of the overall direction – farmers leave the sector as the economy develops, and ultimately those who do tend to be better off for it. But we still know less than we should about the mechanics of adjustment, or how to effect that adjustment smoothly, in a way that optimises food security outcomes.

There are several practical obstacles to a smooth transition. In many, if not most, economies in sub-Saharan Africa, the manufacturing sector has not managed to play a leading role in terms of productive employment generation. Instead, the service sector has been the foremost source of employment creation. For that reason, there has been some focus on the scope for creating more value added within agriculture, and the potential for innovative mixes of large farms and small farms to generate higher incomes (Proctor and Lucchesi, 2012). This search for new development strategies is given impetus by the booming population in Sub-Saharan Africa, where the youth population (aged 15-25) is expected to double over the next 40 years, adding over 40 million to the workforce each ten years.

Box 4.4. Dualism and farm sizes in emerging economy agriculture

The emerging economies included in this study all exhibit dualistic farm structures. Direct comparisons between countries are difficult, because definitions and classification systems vary. However, some general patterns stand out.

In *Brazil*, 84% of holding are categorised as "family" farms, yet these operations occupy just 24% of total agricultural area, the remainder being taken up by "commercial" farms. The average size of a family farm is 18 ha, compared with an average of over 300 ha for commercial farms.

A similar dualism is apparent in *Chile*, where 95% of farms are operated by "individuals", as opposed to corporations, the public sector or communities. However, these farms occupy just 29% of agricultural area (15 million ha), and have an average size of 52 ha. Within this group, "small farms" of less than 12 ha receive specific support. Corporate farms are responsible for a slightly smaller area than individual farms (13 million ha), and have an average size in excess of 1 000 ha.

In *South Africa*, about 80% of agricultural land is occupied by commercial farms, with the remaining 20% farmed by smallholders (a similar breakdown to

---

that in Brazil). However, half the commercial farms earn less than ZAR 300 000 (USD 36 800) per annum, suggesting that most of South Africa's commercial farms are relatively small economic units in international terms. There are about 240 000 small-scale farmers who provide a livelihood to more than 1 million of their family members and occasional employment to another half million. There are approximately another 3 million people in communal farming households that primarily produce for subsistence needs.

In *China*, farm sizes are much smaller. In China, 93% of farms are of less than one ha, while 98% of farms are of less than 2 ha. These small farms account for the vast majority of agricultural area.

In *Indonesia*, production of rice and other food crops is similarly dominated by smallholders with an average area ranging from 0.3 ha in Java to 1.4 ha for irrigated land off-Java. While smallholders are also important suppliers of perennial crops, there are large private and state-owned farms operating mainly in Kalimantan and Sumatra specialised in perennial crops, in particular palm oil and rubber. Their average size is 2 600 ha and they occupy about 15% of the total crop area.

In *Russia*, there are vast numbers of households involved in agricultural production, with operations which have an average size of just 0.4 ha and occupy 5% of agricultural land. Family and peasant holdings occupy 15% of agricultural land and have an average size of 85 ha, while corporate farms occupy 79% of agricultural land and have an average size of over 3 800 ha for "medium and large" operations and 1 164 for "small" ones.

In *Ukraine*, around 70% of total agricultural land and 90% of arable land is owned by individuals. Much of this land is rented out to corporate farms, which have an average size in excess of 2000 ha. In both Russia and Ukraine, smallholders account for about a half of agricultural output, with the other half produced by large-scale operations.

Source: OECD (2011); OECD (2012c).

The broad principle of OECD advice is not to bias incentives, so that the farm household can adjust to where the economic opportunities are greatest. For the majority of farm households in middle income countries, those opportunities will be greater outside farming, and the biggest challenge is to facilitate that transition. But in poorer economies, there may be difficult decisions about when to prioritise agriculture via sector-specific investments (both because of its direct benefits and because of multiplier effects through the rest of the economy) and when to focus on improving opportunities more widely.

4.5. The role for agricultural policies

OECD countries agree on some basic principles with respect to the pursuit of income-related objectives in member countries (OECD, 2002): countries should use social policies to protect incomes in the short term (and provide support for farmers who are unable to adjust), while correcting market failures and investing in public goods in order to strengthen agricultural incomes more fundamentally over the medium to long term. This approach contrasts with using market distorting interventions, such as price supports and input subsidies, which are found to perform poorly in terms of raising the incomes of farm households (OECD, 2001) and often have perverse distributional effects, paying more to larger and richer farmers than to smaller and poorer ones, and taking money away from consumers and taxpayers to boost the incomes for households whose incomes are already above average (OECD, 2003).

In the case of poorer countries, however, it has been argued that this advice might need to be qualified. In the first place, developed country systems of social protection may not be in place and – pending their development – market interventions may be the only practical way of responding to events such as the recent spike in world food prices. Second, market failures are likely to be more endemic, and it may be difficult to tackle them directly. For example, farmers may have low incomes partly because they have no access to credit. Input subsidies have thus been suggested as a practical solution to the otherwise difficult problem of developing input markets and providing financial services to small farmers. Similarly, price stabilisation has been proposed as a relatively simple way of mitigating the impacts of price shocks on poor households, as opposed to market-based forms of risk management or the provision of income safety nets.

In the short-term, price policies provide an easy lever for government, but are inefficient at addressing income concerns. Price support for food products is a blunt instrument because, among the poor, there are net sellers and net buyers of food – in many poor countries, the majority of farm households are net buyers. Price stabilisation (as opposed to price support) can limit the impact of adverse shocks on producers and consumers, but often proves to be fiscally unsustainable. A preferable option for the poor – both producers and consumers – is targeted social programmes, including cash transfers, although these may be difficult to implement in the poorest economies. At the same time, agricultural investments can improve farmers' resilience to risk.

Over the long-term, market interventions treat the symptoms of market failure and underdevelopment, rather than the causes. Price stabilisation can provide a more stable investment climate, but thwarts the development of private risk management, and can export instability onto world markets. Input subsidies can redress failings such as the under-development of infrastructure, missing markets for credit and inputs, and a lack of knowledge of the benefits of using improved seed and fertiliser, but can impede the development of private markets. In both cases, the benefits and costs of intervention need to be judged relative to the benefits and costs of tackling the underlying problems directly.

Finally, there are dangers in using market interventions to address multiple economic and social objectives. Such programmes can become an easy target for interest groups, outliving their original justification and becoming a budgetary millstone. An important priority is that if such policies are to be adopted, then expenditures on them should not crowd out essential investments in support of long-term agricultural development.

4.6.   Risks to food access

In terms of risk and food access, it is helpful to make a distinction between pure consumers and farmers, who are affected by changes on food markets in fundamentally different ways.

Consumers are affected by risk in the agricultural sector predominantly via prices. Indeed, the major policy concern during the food price crisis was with the effect of high prices on consumers budgets (i.e. their real incomes) and the associated implications for food security. Poor consumers spend as much as 50-60% of their incomes on food purchases, and high prices cut deeply into real incomes, causing considerable hardship and potentially longer term humanitarian impacts (FAO, OECD et al., 2011). Price volatility is less of a concern from the consumer side, because consumers commonly benefit from increased volatility around a given price level. This is because food commodities are often substitutes for one another, and changes in the prices of foodstuffs are not perfectly correlated, so food consumers can adjust their food purchases so as to take advantage of price differences (Matthews, 2012b).

In principle governments can use social protection to ensure that price shocks do not have significant implications for consumers' food security. But there has been considerable discussion of the need to intervene by stabilising food prices, for example by imposing export restrictions, suspending tariffs or introducing food subsidies. The shortcomings of such approaches point to the need to build up systems of social protection, so that recourse to less effective policies can be avoided in future. This is especially important if higher and more volatile prices are to become a recurring fact of life.

Market stabilisation is nevertheless seen as having a role to play in developing countries because of (i) market failures, which mean that private markets do not provide adequate risk management; and (ii) under-developed systems of social safety nets. Ultimately, there is a need to address those problems directly, by correcting market failures, for example by improving market information other measures that improving farmers' resilience to – and ability to manage – risk, and by developing effective systems of social protection.

The shortcomings of such approaches point to the need to build up systems of social protection, so that recourse to such policies can be avoided in future. This is especially important if higher and more volatile prices are to become a recurring fact of life. Safety net measures at the international and domestic levels offer more efficient ways of helping producers and consumers cope with food price instability.

Across a range of developed and developing countries, population-wide social safety nets have been used to support the incomes of rural households. In developing countries, conditional cash transfers (CCTs) have become particularly popular over the past decade. These programmes transfer cash to generally poor households on the condition that they make pre-specified investments in the human capital of their children. CCTs have been found to be effective at increasing consumption levels among the poor and have led to behavioural changes, although their impact on final outcomes in health and education has been less clear (Fiszbein and Schady, 2009). This may be due to the need for CCTs to operate in conjunction with complementary investments (e.g. in schools and hospitals). An issue with CCTs is when the "conditional" element is warranted. For example, it may not be worth incurring the monitoring and enforcement costs associated with the condition that parents put their children in school if they would do that anyway.

Emergency food reserves may also be an effective approach to protect the most vulnerable insofar as they can provide subsidised food to specific groups in the community without disrupting private markets. In order to be effective, these reserves need to be combined with an effective early warning system, have transparent and well defined trigger systems, be independent of political processes and be integrated with existing broader social safety nets.

There may also be a specific need for food distribution and nutrition programmes. The widely supported Scale Up Nutrition framework suggests that direct nutrition specific interventions are needed as a complement to a broader multi-sectoral approach to tackle the causes of poor nutrition. Thus they suggest investments to promote good nutritional practices, such as breastfeeding and complementary feeding for infants; interventions to increase the uptake of vitamins, minerals and other micro-nutrients; and therapeutic feeding for malnourished children (SUN, 2012). The wider issues pertaining to nutrition are taken up in Chapter 5 on food utilisation.

A broader, and more complex, set of risks can threaten the food access of producers via their impacts on both prices and production.

A major concern for small farmers is not just the level of prices, but their volatility. Volatile prices pose significant problems for farmers and other agents in food chains, who risk losing their productive investments if those investments require higher prices to be profitable and prices fall after planting decisions have been made (FAO, OECD et al., 2011). Poor smallholders without access to credit may not be able to smooth consumption from one year to the next and may have difficulty financing the crucial inputs needed to plant again. More widely, uncertainty may result in sub-optimal investment decisions in the longer term.

Prices and production are subject to a wide variety of risks (Romer Løvendal and Knowles, 2005). At the global level, those risks include economic risks, such as the recent financial crisis and its implications for trade. At the national level, there are economic and political risks, as well as natural risks such as earthquakes, droughts and floods. Some of those risks may also be felt at the regional or community level, including political strife and natural risks such as landslides, pest attacks and animal diseases. Health risks, which lower production, may be felt at the community level (e.g. HIV/Aids and poor sanitation) and at the individual household level (e.g. illness, injury and disability).

Some risks may be negatively correlated, such as prices and yields in closed markets. Others may have compound impacts – for example households affected by poor health may be more vulnerable to the effects of drought or price shocks. Private agents (individuals, households or communities) seek to reduce their vulnerability to risk via risk management strategies. The effectiveness, or otherwise, of these strategies will determine the role for government policies. As noted in earlier OECD work (OECD, 2000), there is a moral hazard issue in that the more government assumes a role in risk management, the lower the incentives for agents to manage their own risks. As a result, government schemes may "crowd out" private risk management arrangements (OECD, 2009).

Designing risk management tools for farmers is a complex challenge, even in developed OECD countries. In delineating the potential role for governments, OECD's work on risk management has suggested that risks can be divided into three layers (OECD, 2009). First is a "risk retention layer", corresponding to risk that can be effectively managed by farmers and households themselves. Second is an "market layer," which can be addressed by private market instruments such as crop insurance or forward pricing. Third is a "market failure layer", where government intervention may be required. In developing countries, weaker institutions are likely to mean that a wider range of risk falls within the market failure layer, implying a greater need for government action.

Most OECD countries intervene with risk-based policies, either in the form of ex ante risk management (several countries run crop insurance programmes), or ex post protection (such as relief in the event of droughts, natural disasters or other extreme events). Several countries do both. For example, Canada and the United States operate crop insurance programmes and also have disaster programmes.

Developing country farmers – principally smallholders – face additional and more severe risks compared with their counterparts in developed OECD countries. They are typically more vulnerable to natural shocks, emanating from causes such as drought and desertification; and more susceptible to idiosyncratic shocks, such illness and disease. For example, smallholders in sub-Saharan Africa may be more exposed to HIV/Aids, which will affect many aspects of the household's livelihood. The consequences of shocks are also more likely to undermine livelihoods and push households into food insecurity. In countries where there is a lack of effective safety nets, farmers may have to run down their assets, thereby compromising their long-term welfare.

In designing appropriate risk management strategies, the challenge is to discern what can be managed by households, what can be pooled through informal arrangements or formal institutions, what can be insured and marketed, and what can only be addressed through government action. For this, it is important to have an assessment of which risks can be managed by smallholders themselves, and where other agents – community organisations, private sector institutions, national governments and international organisations (including donors) – have a role. A few key aspects of smallholder risk are noted below:

- Informal risk management strategies are widely adopted at both the household and community levels, often as a response to a lack of institutions that help in risk mitigation. For example, insurance and credit markets are usually under-developed. Households can manage farm risk by avoiding certain production risks, inter-cropping, diversifying crop mixes, saving and building stocks; they can manage wider income risk by diversifying their income sources. At the community level, common informal management mechanisms include crop sharing, common property management, mutual social support, informal risk pooling, and rotating savings and credit.

- While agricultural activity can carry income and food security risks, there are also risks from pursuing off-farm opportunities, and maintaining land rights and some agricultural activity can be part of a strategy for smoothing income and consumption. Recent work on Viet Nam illustrates the importance of this strategy (OECD, 2012e).

- Investment can enable households to escape poverty. But that investment may not be forthcoming because of non-functioning credit markets, as a result of constraints on either the supply side or the demand side. On the supply side, banks may be unwilling to lend to smallholders because of asymmetric information, and the costs of obtaining information on creditworthiness. They may also be concerned that poor farmers facing an income shock will refuse to repay loans for the simple reason that doing so in the event of a shock could endanger their immediate food security. On the demand side, households may undertake less investment because the downside risks could threaten their food security. As a result they may choose low risk and low return activities which lead to them "staying secure by staying poor" (Wood, 2003). A safety net policy may help to redress the lack of demand for investment.

A key point is that risk management decisions are inter-related, and outcomes are likely to vary according to a country's structural and institutional characteristics, as well as its general level of development. More generally, there are complex interactions between the risks that smallholders and other farm households face, the risk management strategies they adopt (either directly or indirectly through risk-sharing arrangements), and government policies that affect decisions and outcomes. Governments can intervene in a variety of ways, for example through social protection, targeted transfers, consumer subsidies, support for insurance and reinsurance schemes, and by fostering the development of institutions. Those interventions may "crowd out" farm households' own risk management (Dercon, 2005); but in some cases they may also "crowd in" certain risk management strategies, by alleviating cash constraints.

4.7.   Food access and the role of trade

As noted in Chapter 3, trade openness leads to a different set of relative prices, compared with an environment in which markets are protected. That leads to efficiency gains, but creates winners and losers, the winners being consumers and potential exporters, the losers those producers who formerly benefited from price protection. In a dynamic sense, the producer gains translate into export opportunities, while the losses correspond to adjustment pressure. In terms of food security, opening agricultural markets should lower domestic food prices, leading to gains for consumers. However, there are concerns about the effects on poor smallholder farmers. On one side, there are fears that only larger commercial farmers may be in a position to benefit from improved export opportunities. On the other, are fears that smallholders may not be in a position to compete with lower priced food available on international markets.

Some of the constraints work both ways: if smallholders are constrained from competing in markets by high transport, marketing and distribution costs, then that should also limit their exposure on local markets to import competition. Reducing these costs would enhance their ability to compete in export markets but amplify price pressure from imports. It may also create specific opportunities for smallholders to benefit from access to global markets, in particular for higher value crops. More generally, opening up to trade and facilitating trade by reducing the costs of getting goods to markets can be seen as reforms that cause resources to shift into activities in which they can be more productively engaged, and which complement and accelerate the agricultural transformation described earlier. The implications for incomes and food access are taken up below.

Agri-food exports, smallholders and food security

The relationship between agricultural exports, development and poverty has been central to modern development economics since the 1950s. Myint (1958) argued that the main gain from exports of agricultural products was that it provided a 'vent for surplus' arising from the existence of unemployed resources in developing countries. The existence of overseas markets permitted a country to increase its output and employment by, conceptually, moving from well inside its production possibilities frontier to a point nearer to or on the frontier. He argued that this mechanism helped to explain the rapid growth of production and output of traditional agricultural and primary products in developing countries in the nineteenth century.

More generally, from the perspective of individual producers and smallholders in particular, the creation of markets where none previously existed has the potential to bring gains in output and income which are a multiple of the traditional allocative efficiency gains from participation in trade. A key constraint for smallholder agriculture in developing countries is that farmers practice either subsistence farming or operate largely in local markets due to lack of connectivity to more rewarding markets at provincial, national or global levels. As a result, incentives remain weak, investments remain low, and so does the level of technology adoption and productivity, resulting in a low level equilibrium poverty trap (Torero, 2011). Smallholder access to export markets can help to enhance and diversify the livelihoods of lower-income farm households and reduce rural poverty more generally (World Bank, 2007).

Various studies have shown that access to developed country markets can provide higher revenues for smallholder farmers. In Guatemala, the export of horticultural production generated gross margins per hectare 15 times as large as for maize production, with gross margins per labour day that were twice as large (von Braun and Immink 1994). In Kenya, McCulloch and Ota (2002) found that farm households involved in export horticulture had higher incomes after controlling for farm size, education, irrigation, and other factors compared to farm households that were not involved. However, there is no inevitability that smallholders can take advantage of these opportunities in the absence of public investments to underpin market transactions. In Nepal, despite the clear revenue advantages of cash cropping, farmers have been reluctant to commit to producing for the market given the rudimentary infrastructure and the high variability of prices. As a result, the costs and benefits of developing markets have been unevenly distributed, with smallholders unable to capitalise on market opportunities and wealthier farmers engaging in input intensive cash cropping (Brown and Kennedy, 2005).

These constraints of weak market linkages and high market frictions have been amplified by a new set of challenges associated with compliance with product and process standards (Lee et al., 2012). Sometimes these standards are set and enforced by governments, but increasingly compliance is required even to gain access to private sector supply chains. There is concern that the productivity or production cost advantages that small-scale farmers might have are increasingly outweighed by the escalating transaction costs associated with facilitating, monitoring, and certifying their compliance with standards. There is a risk of a growing polarisation between agribusiness and smallholder farming systems, reducing the poverty alleviating effects of trade if smallholders are excluded or pushed out of high-value production chains as a result (Vorley and Fox, 2004).

More generally, growing corporate concentration in trading, processing, manufacturing and retailing, raises concerns about potential market distortions arising from the absence of competitive markets. Stronger actors in the global supply chain may be able to use their market power to extract more favourable terms from suppliers, leading to the risk that the share of the value created in the food chain accruing to smallholders and processing firms in developing countries declines over time. This can compromise agriculture's potential to act as an effective route for small producers to exit poverty and improve their food security (Vorley and Fox, 2004). These concerns are not unique to international trade. Supermarkets are increasingly important buyers in some developing countries, particularly for the high-value products meeting specific consumer demands related to production process and quality. In Latin America, supermarkets buy 2.5 times more produce from local farmers than the region exports to the rest of the world (Reardon and Berdegué, 2002) and are increasingly important players in Asia and Africa, where smallholder agriculture is concentrated.

There are different answers in the literature as to whether these developments in global food supply chains create a serious barrier to using agricultural exports to alleviate rural poverty and improve food security. Some analyses conclude that modern supply chains lead to the exclusion of small farmers who cannot comply with high food standards (Swinnen, 2007; Reardon et al., 2009). Others highlight the benefits of the new supply chains in providing information on new products, extending input, credit and advice, as well as making marketing services available. These can ease the resource constraints as well as reduce the production and marketing risks that smallholders otherwise face. Demonstrating compliance to high food standards can facilitate access to markets which might otherwise remain closed. Smallholders and rural households can benefit from high-value export production either directly (such as participating through contract farming) or indirectly through the employment created on large scale estate production or agro-industrial processing (Minten et al., 2009; Maertens et al., 2009).

Much of the debate on smallholders and standards has focused on the experience of horticulture with the GLOBLG.A.P approach (Jaffe et al., 2011).[3] Their study takes an agnostic view, arguing that emerging standards are infrequently the primary factor in smallholder market 'exclusion' but also not commonly a primary vehicle for poverty reduction and sustainable smallholder competitiveness. Based on a large survey of African fruit and vegetable exporters, they found evidence for both the optimists and pessimists regarding the prospects for continued smallholder participation in Africa's fresh produce export trade. On the one hand, major buyers often operated indirect procurement chains involving smallholders alongside direct farm-integrated supplies and the majority reported that they planned to maintain or even increase their purchases from smallholders in the future. On the other hand, the survey found evidence that overall numbers of smallholders supplying the main product to the respondent firms had fallen over time, although often the reason for this fall had nothing to do with standards. Jaffe et al. estimate the total number of African smallholders outside South Africa involved in horticulture exports at around 55 000, and suggest that the focus on smallholders and horticulture has been misplaced. They claim that the largest welfare benefits from export-oriented horticulture relate to employment rather than to direct smallholder produce supply. They further conclude that the largest opportunities for future welfare gains from smallholder engagement in markets relate to the development of domestic and regional value chains involving much larger numbers of producers, with consequent benefits also accruing to domestic consumers.

These complexities make clear that there is no inevitability that agricultural exports will necessarily reduce rural poverty and improve food security. Public policy interventions have an important role to help make export markets work for the poor. Public investments in rural transportation and market infrastructure as well as the provision of support services are essential for small farmers to effectively participate in markets and to minimise risk. Helping small farmers to organise through farm associations and co-operatives can assist smallholders overcome diseconomies of scale and bargain more effectively.

Agricultural protectionism and food security

Many developing countries are more focused on the impact of food imports and import competition on household food security, fearing that import competition could undermine the livelihoods of poor food producers. Reflecting this perspective, import protection is advocated to promote greater food self-sufficiency on the grounds that this would not only improve food availability but also lead to greater household food security. The evidence suggests that such policies, on balance, are more likely to undermine access to food for poor consumers, a group which often includes a majority of farmers who are net buyers of food staples, than to increase it. There are more effective policies available to governments to improve households' food access. In particular, governments can stimulate increased food production by investing in measures to sustainably increase food production productivity and to link smallholders better to markets as recommended earlier in this study.

Lower food prices reduce the attractiveness of investing in food production. The pro-poor benefits of agricultural growth in developing countries are now well established (Christiaensen et al., 2011; de Janvry and Sadoulet, 2010). This fact is often used to justify raising food prices through import protection in order to provide incentives for domestic production of staple foods. Paradoxically, however, the main beneficiaries of price support are not smaller food-insecure farm households but rather the more commercial farms with significant food surpluses to sell. More importantly, import protection policies distract attention from the more effective measures governments can take and are likely to hinder the competitiveness of the agricultural sector.

While expanding agricultural market access opens up opportunities to develop the farm sector and to improve the livelihoods of the poor, trade and trade liberalisation also has the potential to disrupt local agricultural markets. The empirical literature shows that the immediate impacts of agricultural trade liberalisation in developing countries for poverty and food security have often been mixed, and depend on the distribution of net buyers and net sellers of staple foods among the poor (see, for example, the studies cited in FAO, 2005). Valdés and Foster (2007), for example, note that reforms in Latin America affected agricultural subsectors in different ways. Producers of exportables generally gained, as did wage-earners in agriculture and processing, whereas small food producers who faced greater import competition often lost in the short run. Overall, however, they conclude that the reforms did not contribute to an increase in rural poverty, and in some cases helped to reduce rural poverty, as in Chile and Colombia.

In the longer-term, these impact effects are outweighed by the adjustments that households make to such terms of trade shocks and are dominated by the impacts on agricultural growth and productivity. Hassine et al. (2010) find strong support for the positive effect of trade openness on agricultural productivity growth through the transfer of technology from more advanced countries based on empirical evidence for 14 Mediterranean countries. Using their empirically-estimated relationship, they conclude that agricultural trade liberalisation in Tunisia would reduce poverty in that country. On the other hand, Yu and Nin-Pratt (2011) in examining the factors behind accelerating total factor productivity in Sub-Saharan Africa in recent years conclude that high dependence on agricultural imports is associated with agricultural productivity slow-downs – this may be because, as noted in Chapter 3, import dependence in these cases is more a result of development failure than of resources shifting successfully into more profitable non-agricultural activities.

More generally, the opportunities and pressures created by international trade are just one dimension of the structural transformation in agriculture and food supply chains documented earlier. Agrifood markets are changing rapidly in many countries with a reduced role for the state, changes in consumer preferences and purchasing power, and the modernisation of food processing and retailing. Enhancing the food security of poor households in this rapidly-changing environment requires a broader focus than just on trade alone and must be seen in the wider context of structural adjustment between the farm and nonfarm economies in developing countries. For many developing countries, the positive food security impacts of trade on non-agricultural incomes, especially jobs and wages, will be the most important contribution of trade.

Notes

1. On the other hand, Sub-Saharan Africa generally performs better that South Asia on a number of anthropometric measures of food insecurity.

2. Lower prices are driven fundamentally by productivity improvements. In a competitive market, the gains from a productivity improvement, i.e. lower unit costs, are whittled away by the entry of new suppliers. However, there are gains to the "early bird" adopters of new technologies (Cochrane, 1958).

3. GLOBALG.A.P. is the world's most widely-used standard for good agricultural practices.

References

Anderson, K. (ed.) (2008), *Distortions to Agricultural Incentives: A Global Perspective, 1955-2007*, Palgrave MacMillan, London and World Bank, Washington, DC.

Benson, T., S. Mugarura and K. Wanda (2008), "Impacts in Uganda of rising global food prices: The role of diversified staples and limited price transmission", *Agricultural Economics*, Vol. 39, Supplement, pp. 513–524.

Bezemer, D and D. Headey (2008), "Agriculture, development and urban bias", *World Development* 34, pp. 1342–1364.

Brown, S. and G. Kennedy (2005), "A case study of cash cropping in Nepal: Poverty alleviation or inequity?", *Agriculture and Human Values* 22 (1), pp. 105–116.

Chen, S. and M. Ravallion (2010), "The developing world is poorer than we thought, but no less successful in the fight against poverty", *Quarterly Journal of Economics*, 125(4), pp. 1577-1625.

Christiaensen, L., L. Demery and J. Kuhl (2011), "The (evolving) role of agriculture in poverty reduction-An empirical perspective", *Journal of Development Economics*, 96 (2), pp. 239–254.

Cochrane, W. (1958), *Farm Prices: Myth and Reality*, University of Minnesota Press, Minneapolis.

Collier, P. and S. Dercon (2009), "African agriculture in 50 years: Smallholder in a rapidly changing world?", Presented at the FAO Expert Meeting on How to Feed the World in 2050, FAO, Rome.

de Janvry, A. and E. Sadoulet (2010), "Agricultural growth and poverty reduction: Additional evidence", *The World Bank Research Observer* 25 (1), pp. 1–20.

Demeke, M., G. Pangrazio and M. Maetz (2009), "Country responses to the food security crisis: Nature and preliminary implications of the policies pursued", *Initiative on Soaring Food Prices*, FAO, Rome.

Dercon, S. (2005), *Insurance Against Poverty*, Oxford University Press, Oxford.

Easterly, W. (2008), "Planners vs searchers in African agricultural aid", *FAO Pro-Poor Livestock Policy Initiative*, FAO, Rome.

Fan, S., B. Yu and A. Saurkar (2008), "Public spending in developing countries: Trends, determination and impact", in *Public Expenditures, Growth and Poverty*, Fan, S. (ed.), John Hopkins University Press, Baltimore.

FAO (2012), *Food Security Indicators*, Revised November, 27, FAO, Rome.

FAO (2005), *The State of Food Insecurity in the World 2005. Eradicating world hunger – key to achieving the Millennium Development Goals*, FAO, Rome.

FAO, OECD et al. [for G20] (2011), "Price volatility in food and agricultural markets: Policy responses", Policy Report including contributions by FAO, IFAD, IMF,OECD, UNCTAD, WFP, the World Bank, the WTO, IFPRI and the UN HLTF, Seoul: G20.

Ferreira, F.H.G., A. Fruttero, P. Leite and L. Lucchetti (2011), "Rising food prices and household welfare. Evidence from Brazil in 2008", *Policy Research Working Paper* No. 5652, The World Bank, Washington, DC.

Filipski, M. and K. Covarrubias (2012), "Distributional impacts of commodity prices in developing countries", in J. Brooks (ed.), *Agricultural Policies for Poverty Reduction*, pp. 61-88, OECD Publishing, Paris.

Fiszbein, A. and N. Schady (2009), *Conditional Cash Transfers: Reducing Present and Future Poverty*, World Bank, Washington, DC.

Hassine, N.B., V. Robichaud and B. Decaluwé (2010), "Agricultural trade liberalization, productivity gain and poverty alleviation: A general equilibrium analysis", *Cahiers de recherche Working Paper* No. 10-22, Centre Interuniversitaire sur le Risque, les Politiques Economiques et l'Emploi, Quebec.

Hazell, P., C. Poulton, S. Wiggins and A. Dorward (2007), "The future of small farms for poverty reduction and growth", *IFPRI 2020 Discussion Paper* No. 42, IFPRI, Washington, DC.

Headey, D. (2013), "The impact of the global food crisis on self-assessed food security", *Policy Research Working Paper* No. 6329, The World Bank, Washington, DC.

IFAD (2010), *Rural Poverty Report 2011*, International Fund for Agricultural Development, Rome.

Irz, X., C. Lin Lin Thirtle and S. Wiggins (2001), "Agricultural productivity growth and poverty alleviation", *Development Policy Review*, 19(4), pp. 449-466.

Ivanic, M. and W. Martin (2008), "Implications of higher global food prices for poverty in low-income countries", *Agricultural Economics*, 39(s1), pp. 405-416.

Jaffe, S., S. Henson and L. Diaz Rios (2011), "Making the grade-smallholder farmers, emerging standards, and development assistance programs in Africa-a research program synthesis", The World Bank, Washington, DC.

Jayne, T.S. et al. (2003), "Smallholder income and land distribution in Africa: Implications for poverty reduction strategies", *Food Policy*, 28(3), pp. 253-275.

Jones, D. and A. Kwiecinski (2010), "Policy responses in emerging economies to international agricultural commodity price surges", *OECD Food, Agriculture and Fisheries Working Papers* No. 34, OECD Publishing, Paris.

Kearney, J. (2010), "Food consumption trends and drivers", *Philosophical transactions of the Royal Society*, 365(1554), pp. 2793–2807.

Lee, J., G. Gereffi and J. Beauvais (2012), "Global value chains and agrifood standards: Challenges and possibilities for smallholders in developing countries", *Proceedings of the National Academy of Sciences*, 109, pp. 12326–12331.

Maertens, M., B. Minten and J. Swinnen (2009), "Growth in high-value export markets in Sub-Saharan Africa and its development implications", *LICOS Discussion Paper Series* No. 245, Katholieke Universiteit Leuven, Leuven.

Matthews, A. (2012b), "Agricultural trade and food security", Background paper prepared for OECD.

McCulloch, N. and M. Ota (2002), "Export horticulture and poverty in Kenya", *Working Paper No. 174*, Institute of Development Studies, Brighton.

Minot, N. (2011), "Transmission of world food price changes to markets in Sub-Saharan Africa", *IFPRI Discussion Papers* No. 1059,IFPRI, Washington, DC.

Minten, B., L. Randrianarison and J.F.M. Swinnen (2009), "Global retail chains and poor farmers: evidence from Madagascar", *World Development*, 37 (11), pp. 1728–1741.

Morris, M., H. Binswanger-Mkhize and D. Byerlee (2009), *Awakening Africa's Sleeping Giant: Prospects for Commercial Agriculture in the Guinea Savannah Zone and Beyond*, The World Bank, Washington, DC.

Myint, H. (1958), "The 'classical theory' of international trade and the underdeveloped countries", *The Economic Journal*, 68 (270), pp. 317–337.

Nelson, G., A. Palazzo, C. Ringler, T. Sulser and M. Batka. (2009), "The role of international trade in climate change adaptation", *ICTSD-IPC Platform on Climate Change, Agriculture and Trade Series Issue Brief* No. 4, ICTSD, Geneva.

OECD (2012a), *Policy Framework for Investment in Agriculture*, OECD Publishing Paris

OECD (2012b), "Policy coherence and Food Security: The effects of OECD countries' agricultural policies", Paper prepared for OECD Global Forum on Agriculture, 26 November 2012, Paris.

OECD (2012c), *OECD Review of Agricultural Policies: Indonesia 2012*, OECD Publishing, Paris.

OECD (2012d), *Agricultural Policies for Poverty Reduction*, OECD Publishing, Paris.

OECD (2012e), "Smallholder risk management in developing countries", OECD Publishing, Paris.

OECD (2011), *Agricultural Policy Monitoring and Evaluation: OECD Countries and Emerging Economies*, OECD Publishing, Paris.

OECD (2009), *The Bioeconomy to 2030: Designing a Policy Agenda, Main Findings and Policy* Conclusions, OECD Publishing, Paris.

OECD (2008), *OECD Employment Outlook 2008*, OECD Publishing, Paris.

OECD (2003), *Farm household income: Issues and policy responses*, OECD Publishing, Paris.

OECD (2002), *Agricultural policies in OECD countries: A positive reform agenda*, OECD Publishing, Paris.

OECD (2001), *Market effects of crop support measures*, OECD Publishing, Paris.

OECD (2000), *Income risk management in agriculture*, OECD Publishing, Paris.

OECD and Eurostat (2005), *Oslo Manual Guidelines for Collecting and Interpreting Innovation Data*, OECD Publishing, Paris.

OECD and FAO (2012), *OECD/FAO Agricultural Outlook 2012-2021*, OECD Publishing, Paris and FAO, Rome.

Proctor, F. and V. Lucchesi (2012), "Small-scale farming and youth in an era of rapid rural change, Knowledge programme small producer agency in the globalised market", IIED, London and HIVOS, The Hague.

Reardon, T. and J.A. Berdegué (2002), "The rapid rise of supermarkets in Latin America: Challenges and opportunities for development", *Development Policy Review*, 20 (4), pp. 371–388.

Robles, M. and M. Torero (2010), "Understanding the impact of high food prices in Latin America", *Economia*, 10(2), pp. 117–164.

Romer Løvendal, C. and M. Knowles (2005), "Tomorrow's hunger: A framework for analysing vulnerability to food insecurity", *ESA Working Paper* No. 05-07, FAO, Rome.

Sen, A. (1984), *Resources, Values and Development*, Basil Blackwell, Oxford.

Sharma, R. (2005), "Overview of reported cases of import surges from the standpoint of analytical content", *FAO Import Surge Project Working Paper* No. 1, FAO, Rome.

SUN (2012), "SUN movement: Revised road map", Secretariat of the Scaling Up Nutrition Movement.

Swinnen, J.F.M. (2007), *Global Supply Chains, Standards and the Poor*, CABI publications, Oxford.

Timmer, P. (2010), "Management of rice reserve stocks in Asia: Analytical issues and country experience", in *Commodity Market Review* 2009-10, pp. 87–120, FAO, Rome.

Timmer, P.C. (1998), "The agricultural transformation", in C.K. Eicher and J.M. Staatz (eds.), *International Agricultural Development*, Johns Hopkins University Press, Baltimore.

Torero, M. (2011), "A framework for linking small farmers to markets", International Fund for Agricultural Development, Rome.

Valdés, A. and W. Foster (2007), "The breadth of policy reforms and the potential gains from agricultural trade liberalization: An ex post look at three Latin American countries", in McCalla, A.F. and J. Nash (eds.), *Reforming Agricultural Trade for Developing Countries, Volume One: Key Issues for a Pro-poor Development Outcome of the Doha Round*, pp. 244-296.

von Braun, J. and M.D.C. Immink (1994), "Nontraditional vegetable crops and food security among smallholder farmers in Guatemala", in J. von Braun and E. Kennedy Agricultural Commercialization, *Economic Development, and Nutrition*, pp. 189-203, John Hopkins University Press, Baltimore.

Vorley, B. and T. Fox. (2004), "Global food chains-Constraints and opportunities for smallholders", Prepared for the OECD DAC POVNET Agriculture and Pro-Poor Growth Task Team, Helsinki Workshop, 17-18 June 2004.

Warr, P. (2008), "World food prices and poverty incidence in a food exporting country: A multihousehold general equilibrium analysis for Thailand", *Agricultural Economics*, 39, (s1), pp. 525-537.

Wiggins, S. (2009), "Can the smallholder model deliver poverty reduction and food security for a rapidly growing population in Africa?", FAO Expert Meeting on How to feed the World in 2050, FAO, Rome.

Wood, G. (2003), "Staying secure, staying poor: The 'Faustian bargain'", *World Development*, 31, 3, pp. 455-71.

World Bank (2007), *World Development Report 2008: Agriculture for Development*, The World Bank, Washington, DC.

Yu, B. and A. Nin Pratt (2011), "Agricultural productivity and policies in Sub-Saharan Africa", *IFPRI Discussion Paper* 01150, International Food Policy Research Institute, Washington, DC.

Chapter 5

Food utilisation and nutritional outcomes

This chapter examines the extent to which income growth explains food security outcomes, and identifies necessary complements such as improved health and sanitation. It also examines the allocation of Official Development Assistance in support of food and nutrition security.

This chapter considers how far income growth goes in contributing to food security and seeks to identify some of the necessary complements. In principle, such information should be able to inform development priorities and targeted interventions to improve nutritional outcomes. The allocation of Official Development Assistance in support of food and nutrition security is then discussed.

5.1. The complements to income growth needed to improve nutrition

A number of other conditions need to be met to ensure consistently adequate nutritional outcomes and the attainment of full food security. The UNICEF framework (depicted in Figure 5.1) sets out the main causes of child malnutrition: insufficient access to food is one of the three channels through which malnutrition can result from poverty, the other being poor water, sanitation and health services, and inadequate maternal and child-care practices. A number of underlying cultural, religious, economic and societal factors affect the extent to which resources at national and domestic levels translate into adequate care practices, good quality water, sanitation and health services as well as food access.

The nutrition and public health literature demonstrates clearly the efficiency and cost-effectiveness of nutrition-specific interventions such as information and education programs targeted both to children and mothers, minerals and food supplements, and the provision of primary health care services, clean water and sanitation infrastructure (Wiggins, 2012; Headey, 2013).

The impacts of agricultural development and economic growth on nutritional outcomes are more difficult to measure. Work undertaken for OECD at the Institute of Development Studies (IDS) reviews the large number of econometric studies that seek to quantify the relationship between income and nutritional outcomes, and to identify the significance and importance of other complements (Masset and Haddad, 2012). A central finding of this meta-evaluation is that the elasticities of nutritional outcomes (such as stunting or underweight) are low with respect to income, and that factors other than incomes are crucial in terms of explaining nutritional outcomes. The implication is that although growth is necessary for food security to progress, growth alone will not be sufficient to accelerate progress on the MDG target of having the prevalence of underweight among under-fives by 2015. Yet, none of the parallel causes of malnutrition are independent of income. The overall level of national income is a determinant of the state's availability to pay for key public services, while individual incomes also determine the household's uptake of education, and its access to health, water and sanitation.[1] The composition of income growth matters as well as the overall rate, as it is increases in the incomes of the poorest that have the greatest impact on nutritional outcomes. Moreover, in many countries, the rural poor are discriminated against in terms of the provision of basic public services.

Figure 5.1. Causes of child malnutrition

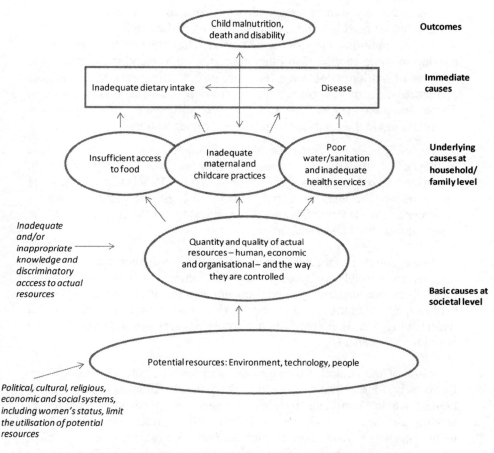

Source: The State of the World's Children 1998 reproduced in Pelletier (2002).

Trade's role in raising incomes was noted in the previous section. In addition, trade can provide consumers with more varied and diversified diets. The positive effects for those currently experiencing under-nutrition have received much less attention in the literature than the potentially negative role of trade in introducing risks of over-nutrition. More generally, Owen and Wu (2007) find that increased openness to trade is associated with lower rates of infant mortality and higher life expectancies, especially in developing countries. Some authors have associated increased trade with a 'nutrition transition' that involves rising rates of obesity and chronic diseases such as cardiovascular disease and cancer (Kearney, 2010). However, these impacts are linked fundamentally to behavioural changes that accompany income growth. Using trade restrictions as a way to modify consumer behaviour is likely to be both inefficient and ineffective.

The relatively weak correspondence across countries between income poverty, under-nutrition and the prevalence of underweight shows how the sufficient conditions for food security and adequate nutrition are clearly missing in some countries. In other countries, conversely, the performance in terms of nutritional outcome (underweight) is strong relative to variable correlated with income growth (poverty) or food access (under-nutrition). These variations across countries suggest differences in development priorities, and in the needs for targeted policy interventions.

Figure 5.2 shows some of these differences. Panel A shows the incidence of underweight, under-nourishment and poverty for eight countries where the incidence of *poverty* is highest. Panel B shows the same indicators for the eight countries with the highest prevalence of *under-nourishment*; while Panel C shows these indicators for the countries with the worst rates of child *underweight*.

Some countries face high levels of undernourishment but less severe rates of underweight and vice versa. For example, in India over 40% of under-fives are underweight, yet undernourishment is below 20%, far behind many countries in this respect (Panel B). Conversely, the rate of undernourishment is almost 50% in Zambia whereas the underweight share is about 15% (Panel C).

Most of the countries where undernourishment is highest are African. Conversely, four of the five countries where the rate of child underweight is highest are in Asia. This paradox has been described among others by Deaton and Drèze (2009) who find that "child under-nutrition is much higher in South Asia than in Sub-Saharan Africa, although the most undernourished countries in both regions fare much the same" (p. 50). On the other hand, de Haen et al. (2011) find that child mortality, which one would expect to be correlated with child malnutrition, is comparatively low in South Asian countries. Looking more broadly across all countries, it is hard to discern regional patterns. Underweight is higher than under-nutrition in about half of African countries; the same is true of Asian countries. In summary, we know less about the causes of these differences than required for effective policy targeting.

Figure 5.2. Underweight, undernourishment and poverty (2004-10)

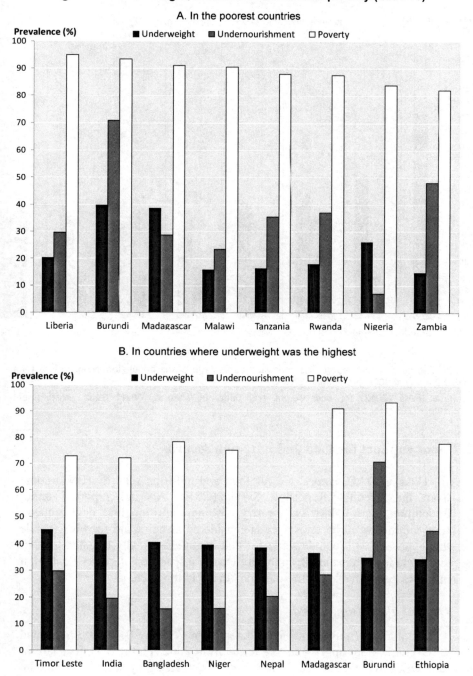

A. In the poorest countries

B. In countries where underweight was the highest

C. In countries where undernourishment was the highest

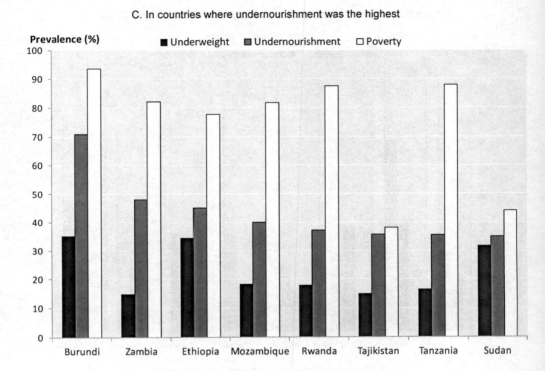

Note: The poverty rate reports the estimated percentage of the population living on less than USD 2 a day at 2005 international prices.

Source: FAO (2012) for underweight and undernourishment, World Bank Development Indicators (2012) for poverty.

5.2. Donor support for food and nutrition security

Data on DAC donors' aid for food and nutrition security (FNS) come from the Creditor Reporting System (CRS). All aid reported under agriculture, agro-industries, forestry, fishing, nutrition and development, food aid/food security assistance is considered as being aid for FNS. While this approach will include some aid that is not specifically targeted to FNS and exclude some which is, in the absence of a specific FNS classification, it provides a reasonable picture of trends in aid in this area.

Overall aid volume trends

Total ODA (multilateral and bilateral) for FNS in 2010 stood at around USD 11.7 billion, up 49% in real terms from 2002 (Figure 5.3). Its share of total ODA over that period, however, has fluctuated only slightly, around an average of 7%. The data show that ODA for FNS has only kept pace with the overall rise in total ODA; there was no evident surge in ODA for FNS following the food price hikes of 2007 and 2008. The share of ODA for FNS began to pick up from a low of 4.5% around 2006 and therefore unrelated to the current interest in FNS.

Figure 5.3. Official Development Assistance for Food and Nutrition Security

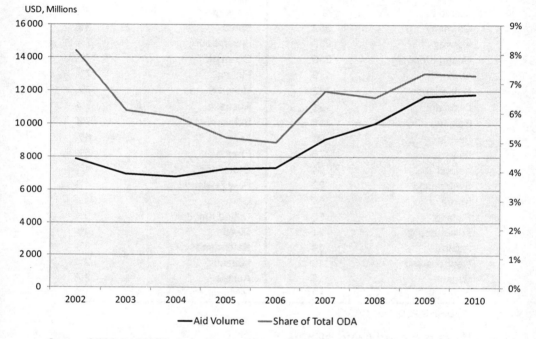

Source: OECD DAC/CRS, Commitments, constant 2010 USD.

Who are the main FNS donors?

In terms of volume, the main FNS bilateral donors are the United States and Japan, which spent on average USD 1.7 billion and USD 1.3 billion p.a. respectively over 2008-10 (Table 5.1). Together, these two donors account for just under half of total bilateral ODA for FNS. In terms of the importance of FNS in a donor country's overall aid programme, countries above the DAC average of 6% for 2008-10 included Canada (12%) as well as Ireland, Japan and Spain (each at 10%).

Table 5.1. Bilateral Official Development Assistance for Food and Nutrition Security: 2008-10 average

Total ODA (MILLION USD)		% OF ODAFOR FNS	
United States	1 708	Canada	12%
Japan	1 364	Ireland	10%
Spain	477	Japan	10%
France	455	Spain	10%
Canada	423	Norway	9%
Germany	352	Korea	8%
Norway	287	Luxembourg	7%
United Kingdom	255	Denmark	7%
Australia	179	Finland	7%
Netherlands	142	United States	7%
Denmark	121	Australia	7%
Belgium	115	Belgium	7%
Sweden	96	Italy	6%
Italy	79	France	5%
Ireland	74	Switzerland	4%
Switzerland	72	New Zealand	4%
Korea	54	Germany	4%
Finland	52	United Kingdom	3%
Luxembourg	21	Sweden	3%
Austria	14	Netherlands	3%
New Zealand	10	Greece	2%
Greece	5	Austria	2%
Portugal	3	Portugal	1%

Source: OECD DAC/CRS.

What is the aid being spent on?

Most ODA for FNS is allocated to agriculture (61% for 2008-10), the second largest category being development food aid at 22%. Compared to 2005-07, there has been little change in the composition of ODA for FNS, despite growing recognition of the persistence and severity of the problem and a better understanding of the comprehensive nature of the causes of FNS, which include but extend well beyond agriculture. ODA for nutrition, for example has remained at 3% of ODA for FNS despite it being increasingly recognised as a critical factor, but this underestimates overall support for nutrition as it does not include sizeable amounts channelled through humanitarian budgets.

Who are the main recipients?

At the regional level, Sub-Saharan Africa received 41% of ODA for FNS in 2009-10 Asia was the other main recipient, with 32%. In terms of income groups, 42% of ODA for FNS went to Least Developed Countries (LDCs) (Figure 5.4). Low Middle Income Countries (LMICs) were the second largest recipient group with 25%. While the share going to Other Low Income Countries (OLICs) appears relatively low (10%), there are in fact only 6 countries in this group now, as compared to 48 LDCs and 40 LMICs.

Figure 5.4. ODA for FNS: Breakdown of geographic and income group 2009-10 average

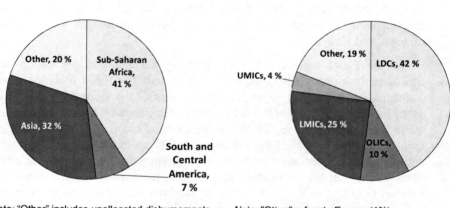

Note: "Other" includes unallocated disbursements.
Source: OECD DAC/CRS, disbursements, current prices.

Note: "Other" refers to Europe (1%), North Africa and Middle East (2%), North America (5%), Oceania (1%), and unallocated (12%).
Source: OECD DAC/CRS, disbursements, current prices.

The top five recipients in 2010 in terms of volume were Afghanistan, Ethiopia, India, Indonesia and Brazil. Compared to 2005, countries such as China and Viet Nam now receive much less ODA for FNS. In terms of ODA for FNS per capita, the top five recipients in 2010 were Afghanistan, Armenia, the West Bank and Gaza Strip, Mali and Bolivia. Overall, there has been considerable movement in ODA for FNS between 2005 and 2010; half of the countries currently in the top 20 recipients list (volume and per capita) were not on the list in 2005, including Ghana, Guinea Bissau, Haiti, Liberia, Mongolia, Niger, Rwanda, Senegal, Sierra Leone and Zimbabwe. In addition, total ODA for FNS in 2010 for both Afghanistan and Mali represents a threefold increase from their 2005 totals.

How much is going to food crisis areas?

Over the past decade, the Horn of Africa and the Sahel have experienced persistent food crises. In the Horn of Africa, total emergency food aid for 2010 amounted to USD 825 million, with ODA for FNS standing at USD 811 million (Ethiopia was by far the main recipient of both emergency food aid (USD 498 million) and ODA for FNS (USD 534 million). In comparison, Chad received USD 139 million in emergency food aid and only USD 47 million in ODA for FNS. The reverse is true for Uganda, which received only USD 31 million in emergency food aid and USD 117 million of ODA for FNS. Looking at ODA for FNS on a per capita basis, again Ethiopia tops the list at USD 6.4 per capita, while Uganda and Eritrea each received less than USD 4 per capita. At present, and triggered by severe under-development, drought and conflict and the resultant massive displacement of people, the areas of highest concern are in Somalia, Ethiopia and Kenya.

In the Sahel region, aid levels both by volume and per capita are considerably higher than in the Horn. Of course, the Sahel includes more countries, including some also classified as part of the Horn of Africa. The Sahel has five countries in the list of top 20 recipients on a per capita basis, while the Horn has none. Total emergency food aid to the Sahel stood at USD 1.3 billion in 2010. Ethiopia and Sudan account for 72% of that total, with Chad and Niger also receiving sizeable amounts. Turning to ODA for FNS, Ethiopia again dominates the picture in terms of volume, accounting for over one-third of the total. However, Burkina Faso, Mali, Niger, Senegal and Sudan were also important recipients in 2010. On a per capita basis, Mali was the main recipient in 2010, with over USD 15 per capita, while Chad, Djibouti, Eritrea, Somalia and Sudan received less than USD 5 per capita. Such low levels of ODA, together with often significant shortfalls in both government and private sector spending, help illustrate why the Los Cabos G20 Summit focused heavily on the pressing challenge of strengthening both emergency and long term responses to food insecurity.

ODA for FNS only represents a portion of the total financing needed to support country FNS plans. ODA supports about one-quarter of the total financing needed, developing country contributions cover another quarter, leaving a financing gap of about 50%. Under the 2003 Maputo Declaration, African countries pledged to spend a minimum of 10% of national budgets on agriculture. Progress towards meeting this target varies considerably across countries, but the average rate across Africa in 2010 was around 6.5%, with only a small number of countries actually meeting or surpassing the target. The most serious problem, however, is finding ways to attract sustainable investment from the private sector (international, domestic and informal). There is still much to do to encourage more private sector investment and public-private partnerships, for example in tackling obstacles related to credit, productivity and risk. This is now a focus of the recent G8 Camp David initiative on the New Alliance for Food Security and Nutrition.

Focusing on trends and patterns of ODA for FNS means focusing on inputs. While this is usually the first part of any aid story, the most important part concerns how well that aid is working. In other words, is it delivering the intended benefits? Considerable work is now underway on two fronts – delivering aid effectively and measuring its results. Based on the DAC Principles of Aid Effectiveness of Accra and Busan and the Rome Principles on Sustainable Global Food Security, donors are increasing support to partner country-owned food security plans and investment strategies, helping to strengthen capacity for in-country implementation and co-ordinating their programmes in partner countries. The Aquila Food Security Initiative (AFSI) group is also currently active in developing a framework for better measuring the results of ODA for FNS, that will cover data collection and common indicators to track progress, and will set out good practices that contribute to the design and implementation of frameworks tailored to the specific situations and needs of partner countries.

Note

1.   The IDS analysis points to shortcomings with many of the studies. First, very few examine the relationship between nutrition status and income with reliable methodologies. Second, the studies are difficult to compare because they often use different nutritional indicators as the dependent variable. Third, the estimates do not consider interaction terms with, for example, food consumption, access to water and sanitation, health care or social protection.

References

de Haen, H., S. Klasen, and M. Qaim (2011), "What do we really know? Metrics for food insecurity and undernutrition", *Discussion Paper* No. 88, Georg-August-Universität Göttingen, Göttingen.

Deaton, A. and J. Drèze (2009), "Food and nutrition in India: facts and interpretations", *Economic and Political Weekly*, 44(7), pp. 42-65.

FAO (2012), *Food Security Indicators*, Revised 27 November, FAO, Rome.

Headey, D. (2013), "The impact of the global food crisis on self-assessed food security", *Policy Research Working Paper* No. 6329, The World Bank, Washington, DC.

Kearney, J. (2010), "Food consumption trends and drivers", *Philosophical transactions of the Royal Society*, 365(1554), pp. 2793–2807.

Masset, E. and L. Haddad (2012), "Income growth and nutrition status: a critical review of the estimated relationship", Background paper prepared for OECD.

Owen, A.L. and S. Wu (2007), "Is Trade Good for Your Health?", *Review of International Economics*, 15, pp. 660–682.

Pelletier, D.L. (2002), "Toward a common understanding of malnutrition. Assessing the contribution of the UNICEF framework", *Background Papers*, World Bank/UNICEF Nutrition Assessment, The World Bank, Washington, DC and UNICEF, New York.

Wiggins, S. and R. Slater (2010), "Food security and nutrition: current and likely future issues", *Science Review* 27, Foresight Project on Global Food and Farming Futures, Government Office for Science, London.

Chapter 6.

Priorities for achieving global food security

This chapter provides the reports main policy conclusions. These fall into three categories: (i) an identification of priorities for global action; (ii) recommendations for ways in policies in OECD can be made more coherent with the goal of global food security; and (iii) broad recommendations in terms of developing countries' own policies.

A well-functioning food and agriculture sector is critical to the attainment of global food security. The sector delivers the food that people eat, while providing a livelihood (including food) to many of the world's poorest. This study has focused on how government policies towards the sector can contribute to the attainment of global food security.

Governments can influence the four dimensions of food security, with policies and investments that increase the **availability** of food sustainably, improve peoples' **access** to it, ensure that their **utilisation** results in adequate nutrition, and guarantee **stability** across those three dimensions.

The principal obstacle to the attainment of global food security is poverty, which constrains peoples' **access** to food. Most of the world's hungry are chronically hungry, and that is because they are poor. The basic requirement for poverty reduction is broad-based development. The underpinnings are mostly well-known but often elusive. They include peace and political stability, sound macroeconomic management, strong institutions, well defined property rights and good governance. The food and agriculture sector has a key role to play in reducing global poverty. More than half of the world's poor depends, either directly or indirectly, on agriculture for their livelihoods. Policies which affect the functioning of the food and agriculture sector have an important role to play in strengthening the incomes of this constituency.

Governments can also increase the **availability** of food via measures that increase supply sustainably or restrain demands that do not translate into improved food security outcomes. The analysis in this study reveals that there is great scope for fundamentally altering supply conditions by raising productivity growth, improving the efficiency of natural resource use, reducing post-harvest losses and adapting to climate change. Equally, changes on the demand side, including reduced over consumption and less consumer waste, could substantially ease the supply side challenge. Because of the wide scope for change in each area, there is a danger of looking for a "magic bullet" in one area that makes actions in the other areas unimportant. However, actions are needed across all the areas discussed in the study.

The chief requirements to improve the **utilisation** of food are complementary policies. Improvements in education and primary healthcare can strengthen income growth, and – along with other investments, notably in sanitation and clean water – improve nutritional outcomes. Direct nutrition interventions have also been shown to be effective. However, a well functioning food and agriculture system which improves availability and access (and guarantees their stability) should also increase energy consumption and, with increased diversity of diets, nutrition too.

The fourth way in which policies related to food and agriculture can improve food security is by ensuring **stability**, such that farmers' incomes and consumers' ability to buy food are resilient to shocks. This means helping the food insecure manage domestic risks (such as weather-related risks in the case of farmers) and international risks (such as extreme price swings and trade interruptions).

The four channels are inter-connected, with policies having complementary effects. As an example, policies which raise agricultural productivity strengthen the incomes of farmers and rural communities and with it their food access. They also increase food availability, benefiting consumers (and increasing their access) to the extent that domestic prices are lower than they would otherwise be.[1] They can contribute to reduced income and price risk, ensuring greater stability of access for producers and consumers. Finally, by raising the real incomes of both producers and consumers they may lead to healthier diets and improved utilisation.

Many of the required policies are compatible with the sustainable use of natural resources. Yet while there is important scope for sustainable intensification, current production patterns may not always be compatible with sustainable resource use. In many countries and regions, there is no effective pricing of natural capital, with the result that production is too intensive or occurs in areas where ultimately it should not. The need to price natural capital in order to ensure sustainable resource use is a countervailing force that may, in some circumstances, put upward pressure on food prices. This underscores the primary importance of income growth. Only if incomes grow sufficiently can trade-offs between immediate food security and sustainable resource use be avoided.

Across the four channels through which food security can be enhanced, it is possible to group the policy implications contained in this study into three categories:

- **Identification of priorities for global action.** While some policies can be implemented at the national level, in many areas there are clear gains from multilateral action. In particular, the benefits of widespread trade openness exceed the benefits from unilateral liberalisation. Similarly, multilateral platforms can be an important vehicle for knowledge sharing (for example, in research and development, or in the design of risk management tools).

- **Recommendations for OECD countries.** These consist of specific contributions to global food security that OECD countries can make via reforms to their own policies, in terms of avoiding policies that create negative spill-overs and adopting beneficial policies. The latter includes

sharing knowledge that can be of help to developin⁄
corresponds to the Policy Coherence for Developmer

- **Suggestions on how developing countries car
  food security policies**. National govern⁄
  responsibility for implementing strategies a⁄
  security. The analysis here seeks to clari⁄
  policies needed to ensure food security
  country's level of economic d⁄
  circumstances, including its com⁄
  activities.

6.1.   Needs for global policy acti⁄

High and volatile food ⁄
security more difficult in th⁄
the need for a number of ⁄
world food markets. S⁄
policy report that OE⁄
to the French Pres⁄
Markets: Policy F⁄

One was
co-ordinatic⁄
responses ⁄
Agricul⁄
Response ⁄
by improvin⁄
developments. A.⁄
representatives of o⁄
response to the IO repoɪ⁄
work of Finance Ministers ⁄
transparency and functioning of ⁄

It was also suggested that develop⁄
would be of value to the most vulnerable c⁄
that could include systems of strategically plac⁄
World Food Programme, supported by other IOs,⁄
proposal for a pilot programme for small targeteu⁄
reserves in West Africa.

However, recent high food prices are not just a one-off shoc⁄
appear to reflect a basic structural change that has taken place in wo⁄
markets. While it is hazardous to project food prices, it would appea⁄
unlikely that prices will return to their historic lows in the mediu⁄

6.2.   Policy recommendations for OECD countries

OECD countries can improve global food security by eliminating policies that create negative international spill-overs. The traditional concern is with protection and trade-distorting domestic support, which have the potential to undercut farmers' livelihoods in developing countries. Notwithstanding trade preferences granted to some developing countries, tariffs on agricultural products remain several times higher than those levie⁄ on industrial goods, which restricts market access for developing cou⁄ farmers with export potential. Higher prices have historically led ⁄ accumulation of surpluses, which have been disposed of with the ⁄ export subsidies. These depress international prices, making condi⁄ difficult for competitors on international markets and for impor⁄ producers on domestic markets. Policies to support farmers h⁄ been counter-cyclical, which stabilises domestic marke⁄ instability onto world markets.

These concerns persist, but while agricultural tarif⁄ tariffs in other sectors, OECD countries have on avera⁄ of support that they provide to agriculture, and in s⁄ been a significant re-structuring of policies, with ⁄ decoupled from production decisions. As a re⁄ that support on developing countries are no⁄ have been facilitated in recent years by ⁄ have reduced the gaps between domesti⁄ Moreover, as price gaps have narrowe⁄ domestic support programmes has de⁄ countries have instituted supports⁄ reverse tendency of making inter⁄ otherwise be, while (in the ca⁄ creating a demand that is les⁄ tariff peaks and cases of tar⁄

In the context of hi⁄
policies which create ⁄
where necessary, m⁄
manage risk and ⁄
same time, the⁄
sector's nega⁄
for public ⁄
at main⁄

would add modestly to the level of food prices; but that would be a one-off effect, the rise would be small compared with the recent changes witnessed on world food markets, and the elimination of the policies' counter-cyclical elements would help stabilise world food prices.

Besides avoiding harmful policies, there are many positive ways in which OECD countries can contribute to global food security, in particular by easing the conditions of food availability. Sustainable increases in supply, which can be achieved through productivity increases, are one way of doing that. The returns to public (and private) investment in agricultural research and development are very high, although the lag times are long. Renewed efforts at the national level, accompanied by greater international collaboration, are warranted. At the same time, incentives to encourage more efficient use of land, water and biodiversity resources would contribute to sustainable supply increases in many regions. Innovation, broadly defined to include not just science but education, training, and organisational improvements, also offers a strong potential to mitigate and adapt to the negative impacts of climate change. On the demand side, improved information and public awareness could substantially reduce overconsumption, cut down on consumer waste and facilitate healthy food choices.

The other area for action is in knowledge sharing. OECD countries, in particular countries that have developed recently, have potentially important experiences to share, including with respect to the role that agricultural development has to play in poverty reduction, and in terms of institutional changes and policies that have been effective. There may also be specific knowledge and expertise that can be transferred in areas such as agricultural research and innovation, and farm management techniques. Of course, knowledge sharing works in multiple directions. OECD countries can learn from the experiences of developing countries, and the benefits of information exchange among developing countries are becoming increasingly apparent. The OECD provides mechanisms for policy dialogue so that countries can benefit from such mutual learning.

Overseas Development Assistance has an important role to play in improving food security in some countries, particularly those that do not generate enough tax revenues to pay for essential public investments and services. There is renewed recognition that aid needs to refocus on agricultural development, including promoting agricultural trade, as the sector is a key area of comparative advantage in many developing countries. It is beyond the scope of this study to provide conclusions for aid policies, beyond the principle that allocations should support national strategies. With regard to those strategies, OECD analysis suggests that agricultural development can best be achieved by prioritising agriculture's enabling environment, rather than supporting specific production activities (OECD, 2012). The basic pre-requisites are long term investments in public goods which improve competitiveness, such as research and development and rural infrastructure, coupled with targeted assistance to poorer households via social programmes. Aid for Trade has an important role in improving developing countries' supply capacity, so that they can respond to improved export opportunities.

6.3. Developing countries' own policies

A core message of this study is that the overriding priority for ensuring global food security is to raise the incomes of the poor, and with it their access to food. Agricultural development has a crucial role to play, given that the majority of the world's poor lives in rural areas, where agriculture is the foremost economic activity. But this is not a separate or independent role: agriculture needs to be integrated into wider growth and development strategies. The countries that have been most successful in reducing rural poverty and food insecurity have been the ones in which balanced rural development has allowed a progressive integration of rural and urban labour markets.

Balanced rural development involves promoting agricultural development on the one hand, while broadening opportunities for the many farmers who will have better long-term (i.e. inter-generational) prospects outside the sector. Even with higher prices and greater opportunities within agriculture than there have been for decades, resource-poor farmers will face adjustment pressures and, as incomes rise, the majority of children from farm families will have better prospects outside the sector. The key to striking the right balance is to avoid creating incentives that prejudice the individual's decision on whether to exploit improved opportunities within or outside farming. Focusing exclusively on supporting smallholder structures could trap households into livelihood patterns that – even if they can improve their immediate food security – impede their long-term prospects.

An important challenge, therefore, is to promote an efficient farming structure that is capable of yielding incomes that are comparable with those in the rest of the economy, and doing so in ways that are environmentally sustainable. In many countries, smallholders have a key role to play, because they constitute the dominant type of farm structure. Yet they are often poor and food insecure. In some contexts the immediate priority may be to raise smallholders' incomes directly by investing in smallholder productivity; in other cases it may be more effective to concentrate on building alternative opportunities in the rural economy and beyond.

In many cases, the foremost need is to redress urban bias, which results in under-provision of public goods and essential services, such as health, education, and physical infrastructure (including ICT) in rural areas. Public investments, and public-private partnerships, to provide strategic public goods or quasi public goods for further agriculture development, such as adapted research, training and extension services, are likely to be much more effective over the long term than market interventions, for example through price supports and input subsidies. Even in the short term, with appropriate skills and supporting infrastructure, wider application of already available technologies could help reduce the productivity gap in developing country agriculture, bringing with it significant economic benefits.

There is a particular need for risk-management tools tailored to the needs of vulnerable farmers, which can reduce the effects of price volatility and enable them to manage risks from weather, climate change, pests, macroeconomic and other shocks. At the same time, governments may need to manage a range of national risks, including those emanating from global markets. The development of such tools is being supported by the Platform on Agricultural Risk Management (PARM).

Income growth is essential, and while many countries are making progress, others – mostly located in Africa and South Asia – are being left behind. Moreover, as FAO's latest State of Food Insecurity (SOFI) indicates, income growth is necessary but not sufficient to accelerate reduction of hunger and malnutrition (FAO, 2012a). The composition of growth matters, as more equal growth is likely to lead to faster improvements in the food security of the poorest. So too does provision of the complements necessary for improved nutritional outcomes. These include improved opportunities for the poor to diversify their diets; access to safe drinking water and sanitation; access to health services; better consumer awareness regarding adequate nutrition and child care practices; and targeted distribution of supplements in situations of acute micronutrient deficiencies. Good nutrition, in turn, supports economic growth.

For larger developing countries, it is important to note that their agricultural and associated trade policies have increasingly important impacts in world markets. Indeed, it is no longer relevant, given the changing structure of world trade, to view the spill-over effects of agricultural policies as exclusively an OECD country issue. During the 2007-08 food price crisis, export restrictions were used predominantly by emerging and developing countries, and exacerbated the crisis – as well as placing a specific burden on some developing countries which could not source imports. The use of alternative non trade-distorting policies would provide domestic benefits and avoid undermining other countries' food security.

There is a general tendency to view the food security implications of biofuels solely in terms of their impacts on world food markets. But for a number of developing countries, biofuels could provide important economic opportunities. The realisation of those opportunities could require significant farm level adjustment, with larger operations and relatively more people earning the income from wage labour as opposed to relying on their own food production for their livelihoods. Insofar as that adjustment occurs, the terms under which farmers relinquish their land and the conditions of salaried employment will be an important determinant of the food security implications.

A common theme of this study has been the inter-connectedness between the multiple determinants of food security, and the links between policies that are adopted at national, regional and multilateral levels. The main impetus for improvements in countries' food security will come from their own strategies and policies. But progress at the national level can be supported by improved co-ordination and coherence at the multilateral level; through knowledge sharing in technical areas such as research as well as on best policy practices; and through the catalytic role of aid. It is in these areas that OECD can support countries' efforts to ensure the food security of their citizens.

Note

1. In closed economies the price effect is direct; in open economies the price effect comes via the cumulative impact of all countries' policies on international markets.

References

FAO (2012a), *The State of Food Insecurity in the World*, FAO, Rome.

FAO, OECD et al. [for G20] (2012), "Sustainable agricultural productivity growth and bridging the gap for small-family farms", Interagency Report to the Mexican G20 Presidency with Contributions by: Bioversity, CGIAR Consortium, FAO, IFAD, IFPRI, IICA, OECD, UNCTAD, UN, WFP, World Bank, WTO, G20 Mexican Presidency.

FAO, OECD et al. [for G20] (2011), "Price volatility in food and agricultural markets: Policy responses", Policy Report including contributions by FAO, IFAD, IMF,OECD, UNCTAD, WFP, the World Bank, the WTO, IFPRI and the UN HLTF, Seoul: G20.

OECD (2012), "Policy coherence and Food Security: The effects of OECD countries' agricultural policies", Paper prepared for OECD Global Forum on Agriculture, 26 November 2012, Paris.

ORGANISATION FOR ECONOMIC CO-OPERATION AND DEVELOPMENT

The OECD is a unique forum where governments work together to address the economic, social and environmental challenges of globalisation. The OECD is also at the forefront of efforts to understand and to help governments respond to new developments and concerns, such as corporate governance, the information economy and the challenges of an ageing population. The Organisation provides a setting where governments can compare policy experiences, seek answers to common problems, identify good practice and work to co-ordinate domestic and international policies.

The OECD member countries are: Australia, Austria, Belgium, Canada, Chile, the Czech Republic, Denmark, Estonia, Finland, France, Germany, Greece, Hungary, Iceland, Ireland, Israel, Italy, Japan, Korea, Luxembourg, Mexico, the Netherlands, New Zealand, Norway, Poland, Portugal, the Slovak Republic, Slovenia, Spain, Sweden, Switzerland, Turkey, the United Kingdom and the United States. The European Union takes part in the work of the OECD.

OECD Publishing disseminates widely the results of the Organisation's statistics gathering and research on economic, social and environmental issues, as well as the conventions, guidelines and standards agreed by its members.

OECD PUBLISHING, 2, rue André-Pascal, 75775 PARIS CEDEX 16
(51 2013 05 1 P) ISBN 978-92-64-19534-9 – No. 60741 2013-05